ANTARCTIC RESEARCH SERIES

American Geophysical Union

ANTARCTIC RESEARCH SERIES

American Geophysical Union

Volume 1 BIOLOGY OF THE ANTARCTIC SEAS
Milton O. Lee, *Editor*

Volume 2 ANTARCTIC SNOW AND ICE STUDIES
Malcom Mellor, *Editor*

Volume 3 POLYCHAETA ERRANTIA OF ANTARCTICA
Olga Hartman

Volume 4 GEOMAGNETISM AND AERONOMY
A. H. Waynick, *Editor*

Volume 5 BIOLOGY OF THE ANTARCTIC SEAS II
George A. Llano, *Editor*

Volume 6 GEOLOGY AND PALEONTOLOGY OF THE ANTARCTIC
Jarvis B. Hadley, *Editor*

Volume 7 POLYCHAETA MYZOSTOMIDAE AND SEDENTARIA OF ANTARCTICA
Olga Hartman

Volume 8 ANTARCTIC SOILS AND SOIL FORMING PROCESSES
J. C. F. Tedrow, *Editor*

Volume 9 STUDIES IN ANTARCTIC METEOROLOGY
Morton J. Rubin, *Editor*

Volume 10 ENTOMOLOGY OF ANTARCTICA
J. Linsley Gressitt, *Editor*

Volume 11 BIOLOGY OF THE ANTARCTIC SEAS III
Waldo L. Schmitt and George A. Llano, *Editors*

Volume 12 ANTARCTIC BIRD STUDIES
Oliver L. Austin, Jr., *Editor*

Volume 13 ANTARCTIC ASCIDIACEA
Patricia Kott

Volume 14 ANTARCTIC CIRRIPEDIA
William A. Newman and Arnold Ross

Volume 15 ANTARCTIC OCEANOLOGY
Joseph L. Reid, *Editor*

Volume 16 ANTARCTIC SNOW AND ICE STUDIES II
A. P. Crary, *Editor*

Volume 17 BIOLOGY OF THE ANTARCTIC SEAS IV
George A. Llano and I. Eugene Wallen, *Editors*

Volume 18 ANTARCTIC PINNIPEDIA
William Henry Burt, *Editor*

Volume 19 ANTARCTIC OCEANOLOGY II: THE AUSTRALIAN–NEW ZEALAND SECTOR
Dennis E. Hayes, *Editor*

Volume 20 ANTARCTIC TERRESTRIAL BIOLOGY
George A. Llano, *Editor*

Volume 21 RECENT ANTARCTIC AND SUBANTARCTIC BRACHIOPODS
Merrill W. Foster

Volume 22 HUMAN ADAPTABILITY TO ANTARCTIC CONDITIONS
E. K. Eric Gunderson, *Editor*

Volume 22 | ANTARCTIC RESEARCH SERIES

Human Adaptability to Antarctic Conditions

E. K. Eric Gunderson, *Editor*

Volume 22 | ANTARCTIC RESEARCH SERIES

HUMAN ADAPTABILITY TO ANTARCTIC CONDITIONS

E. K. Eric Gunderson, *Editor*

1707 L Street, N.W.
Washington, D. C. 20036

Library of Congress Cataloging in Publication Data

Gunderson, Ellsworth K. Eric, 1923–
Human adaptability to Antarctic conditions.

(Antarctic research series, v. 22)
Includes bibliographies.
1. Cold adaptation. 2. Cold—Physiological effect.
I. Title. II. Series: American Geophysical Union.
Antarctic research series, v. 22.
QP82.2.C6G86 612'.01446 74-18498
ISBN 0-87590-121-2

List Price $13.00

Published by the
American Geophysical Union
With the aid of a grant from the
National Science Foundation

Printed by
Heffernan Press
Worcester, Massachusetts

THE ANTARCTIC RESEARCH SERIES

The Antarctic Research Series is designed to provide a medium for presenting authoritative reports on the extensive and detailed scientific research work being carried out in Antarctica. The series has been successful in eliciting contributions from leading research scientists engaged in antarctic investigations; it seeks to maintain high scientific and publication standards. The scientific editor for each volume is chosen from among recognized authorities in the discipline or theme it represents, as are the reviewers on whom the editor relies for advice.

Beginning with the scientific investigations carried out during the International Geophysical Year, reports of research results appearing in this series represent original contributions too lengthy or otherwise inappropriate for publication in the standard journals. In some cases an entire volume is devoted to a monograph. The material published is directed not only to specialists actively engaged in the work but to graduate students, to scientists in closely related fields, and to interested laymen versed in the biological and the physical sciences. Many of the earlier volumes are cohesive collections of papers grouped around a central theme. Future volumes may concern themselves with regional as well as disciplinary aspects or with a comparison of antarctic phenomena with those of other regions of the globe. But the central theme of Antarctica will dominate.

In a sense, the series continues the tradition dating from the earliest days of geographic exploration and scientific expeditions—the tradition of the expeditionary volumes that set forth in detail everything that was seen and studied. This tradition is not necessarily outmoded, but in much of the present scientific work, one expedition blends into the next, and it is no longer scientifically meaningful to separate them arbitrarily. Antarctic research has a large degree of coherence; it deserves the modern counterpart of the expeditionary volumes of past decades and centuries that the Antarctic Research Series provides.

With the aid of a grant from the National Science Foundation in 1962, the American Geophysical Union initiated the Antarctic Research Series and appointed a Board of Associate Editors to implement it. Supplemental grants, the income from the sale of volumes in the series, and income from reprints and other sources have enabled AGU to continue this series. The response of the scientific community and the favorable comments of reviewers cause the board to look forward to the continued success of this endeavor.

With the next volume the format and distribution of books in this series will be changed in an effort to reduce delays in disseminating the information presented. The individual papers of a planned volume will be printed and issued as quickly as production schedules will allow. Usually, a paper will be published individually with its own cover; however, if the publication of two or three short papers can be scheduled very close together, they will be distributed together. Papers may be purchased as they are released or by a standing order for the series. The purchaser may then have the papers bound into a volume if he chooses to do so. When the last paper of a volume has been released, cover pages and the table of contents will be issued. It is believed that this method of production and distribution will increase the value of the series for its users by substantially decreasing the time for publication.

To represent the broad scientific nature of the series, the members of the Board were chosen from all fields of antarctic research. At the present time they include Charles R. Bentley, solid earth geophysics; Avery A. Drake, Jr., geology; Worth D. Nowlin, Jr., oceanography; Bruce Parker, biology; David L. Pawson, biology; Martin A. Pomerantz, aeronomy and geomagnetism; Morton J. Rubin, meteorology; Waldo L. Schmidt, member emeritus; and Laurence M. Gould, honorary chairman. Fred G. Alberts, secretary to the U.S. Advisory Committee on Antarctic Names, gives valuable assistance in verifying place names, locations, and maps.

MORTON J. RUBIN
Chairman, Board of Associate Editors
Antarctic Research Series

PREFACE

This volume was made possible by the interest and support of George A. Llano, Office of Polar Programs, National Science Foundation. He originally suggested the need for such a review of research on man's adaptation to the antarctic environment and gave encouragement and assistance in carrying out this objective.

In the planning of this volume the decision was made to solicit contributions from investigators who were actively engaged in antarctic research. An effort was made to encompass all major medical and behavioral investigations and to reflect the multidisciplinary and multinational character of research in Antarctica. It was possible to obtain contributions from distinguished scientists who themselves directed major national programs and who were in excellent positions to review progress over the past decade.

In 1949, Otto G. Edholm founded the Division of Human Physiology, National Institute for Medical Research, where experimental climatic chambers were built and studies were carried out on the physiological effects of the environment on man. His former appointments include Professor of Physiology, University of Western Ontario, where work was begun on the monograph *Man in a Cold Environment* (A. C. Burton and O. G. Edholm, Edward Arnold, London, 1955). Members of the Division of Human Physiology have been engaged in field studies in many parts of the world, including the Antarctic and the Arctic and the Himalayas and Mount Everest, and have collaborated closely with the armed forces on problems of survival in both heat and cold. For many years O. G. Edholm has supervised the training of medical officers for the British Antarctic Survey to undertake a variety of research studies during their tours in the Antarctic. This program has provided continuity of research efforts and close coordination between laboratory and field. He recently organized the first international symposium on human biology and medicine in the Antarctic, which was held at the Scott Polar Research Institute, Cambridge, England; the proceedings have been published in *Polar Human Biology,* edited by O. G. Edholm and E. K. E. Gunderson, Heinemann Medical Books, London, 1973.

Originally, it was planned that Graham M. Budd would assist Edholm in preparation of a chapter encompassing British Antarctic Survey and Australian National Antarctic Research Expedition activities. Budd's contribution was such an extensive and important one, however, that it was incorporated as a separate chapter in the monograph. He wintered as medical officer at Mawson station and at Heard Island and has maintained an interest and involvement in polar physiological studies. He is presently affiliated with the School of Public Health and Tropical Medicine, University of Sydney, Sydney, Australia. Recent medical investigative activities have taken him to the jungles of New Guinea.

Since 1950, Jean Rivolier has conducted extensive research on extreme environments (high elevation, cold, survival) and on phenomena of biometeorological adaptation. Recent work has centered on long-term biological rhythms, sleep, and the psychology of isolated groups. He wintered as medical officer at Géologie Archipelago in 1952 for the French Polar Expeditions. He has written a delightful book on the habits of emperor penguins. He is currently Chief of Medicine for Overseas Territories and the French Antarctic and for French Polar Expeditions, and he is affiliated with the French National Center of Scientific Research.

Jay T. Shurley and his associates initiated a new era in antarctic biomedical research when they established a sophisticated sleep and psychophysiological research program at South Pole station during 1967–1968. He pioneered in studies of sensory, perceptual, and social isolation and has long been interested in the psychophysiology of sleep. He has been the principal investigator for National Science Foundation supported studies of sleeping and dreaming patterns and psychophysiological adaptation in antarctic personnel. Former appointments include psychiatric research administrator at the Clinical Center, National Institutes of Health, Bethesda, Maryland; Professor of Psychiatry and Behavioral Sciences, University of Oklahoma Medical Center; and Chief, Psychiatry Service, Veterans Administration Hospital, Oklahoma City. Current appointments include Senior Medical Investigator, Veterans Administration; Career Research Professor of Psychiatry, University of Oklahoma; and Adjunct Professor of Human Ecology, College of Health, Oklahoma University Health Sciences Center. He currently serves as member, Committee on Polar Research, National Academy of Sciences–National Research Council, and Chairman, Panel on Human Biology and Medicine, and as U.S. member, Working Group on Human Biology and Medicine, Scientific Committee on Antarctic Research.

Kirmach Natani studied engineering and physics at the University of California, Berkeley, and served with the Peace Corps in Thailand. He wintered at South Pole station in 1967 and collected electrophysiological and behavioral data under J. T. Shurley's supervision. These data were the basis for a master's thesis in psychology at the University of Oklahoma. Currently, he is a doctoral candidate in biological psychology, University of Oklahoma Health Sciences Center, and is a research associate in the Behavioral Sciences Laboratories of the Oklahoma City Veterans Administration Hospital.

The editor of the monograph has been engaged in a sustained program of psychological studies in the Antarctic since 1961. This effort, primarily designed to aid in personnel selection for the U.S. program but not limited to this objective, has been supported by the U.S. Navy Bureau of Medicine and Surgery. Current appointments include Head, Epidemiology and Operational Psychiatry Division, Navy Medical Neuropsychiatric Research Unit, San Diego, California; Adjunct Professor of Psychiatry, School of Medicine, University of California, San Diego; and member, Panel on Human Biology and Medicine, Committee on Polar Research, National Academy of Sciences–National Research Council.

E. K. ERIC GUNDERSON

CONTENTS

INTRODUCTION

E. K. ERIC GUNDERSON

Epidemiology and Operational Psychiatry Division, U.S. Navy Medical Neuropsychiatric Research Unit San Diego, California 92152

SCOPE AND LIMITATIONS

This volume reviews research in human biology, physiology, medicine, psychiatry, and psychology conducted in the southern polar regions during the past decade, and primary attention is given to recent studies. Earlier studies are described where such description is needed to give historical perspective. Generally, the research to be summarized is concerned with human subjects who spent an entire year or more in the antarctic region.

The study of man's adaptation to the antarctic environment within a comprehensive scientific framework is a recent development. It is evident from this review that useful beginnings have been made, and a wide range of topics is represented in the present volume. Measurement of biochemical, physiological, and psychological states and analyses of their interrelationships as well as their correlations with environmental variables present a formidable challenge, even under the most carefully controlled laboratory conditions and with the most advanced equipment. To attempt such studies under extreme field conditions would seem to involve staggering and perhaps insurmountable difficulties; nevertheless, essential groundwork has been laid for more specialized and sophisticated types of investigations in the future. The integration of laboratory research and field investigations in this setting would seem to have a great deal of value for human biology.

EARLY STUDIES

One of the first groups of men to endure a winter in the Antarctic was the crew of the *Belgica,* carrying a Belgian scientific party including Roald Amundsen as first mate. The physician on the *Belgica* was Frederick A. Cook, who gave a detailed account of the party's struggle for survival during 13 months when it was locked in the ice. The men suffered from scurvy and exhaustion, but even more they suffered from homesickness, mental disturbances, and boredom. In his journal *Through the First Antarctic Night,* Cook recorded the incredible hardships of the terrible winter and taught future antarctic explorers much about the need for proper clothing, food, shelter, and equipment as well as about the demoralizing effects of inactivity and boredom.

The next year, 1899, C. E. Borchgrevink and nine companions were the first men to winter on the antarctic continent at Cape Adare. The continent was not continuously occupied until near the middle of the century. In 1956–1958 a gigantic invasion of the continent took place during the antarctic phase of the International Geophysical Year (IGY). Twelve nations built more than 40 stations on the mainland and surrounding islands. Occupation of some of these bases has continued to the present.

Physicians played prominent roles in early antarctic explorations, and a number wrote extensive accounts of their heroic and often tragic adventures [*Silverberg,* 1965]. Natural science observations made during early pioneering expeditions included general accounts of health problems by medical officers, but few data of scientific value were obtained.

Admiral Byrd's account of his several months alone at an antarctic weather station gives a vivid picture of the extreme stresses faced by an individual in complete isolation. Byrd felt that living in solitude was preferable to living in isolation with another person, however, because of the inevitable conflicts that would arise [*Byrd,* 1938].

Planned medical investigations in wintering parties began with Byrd's expeditions and with the Norwegian-British-Swedish Antarctic Expedition that spent two winters at Maudheim during 1949–1952. The scientific programs of the IGY did not officially include biology or medicine, although a few studies in these areas were conducted on the initiative of interested individuals or agencies.

Much of the systematic physiological research conducted in the Antarctic has been under the auspices of the British Antarctic Survey. Recent results of this program are described by O. G. Edholm in the present volume. Early studies in this series were previously summarized in detail by *Edholm and Lewis* [1964] and *Edholm* [1965].

Wilson [1965] provided a comprehensive review of physiological and medical studies before 1963. These early physiological studies were concerned primarily with cold adaptation, diurnal rhythms, work loads and food consumption, metabolic and hormonal changes, cardiovascular functions, and relationships between meteorological weather, microclimate, and skin temperature. Measurements often were made under difficult conditions, and generally only crude instruments were available. Relatively few studies of psychological adjustment in wintering parties were available at that time.

Medical problems and health programs in the polar regions were reviewed at two international conferences. The Conference on Medicine and Public Health in the Arctic and Antarctic, sponsored by the World Health Organization, was held in Geneva, Switzerland, in August 1962. This conference was an exploratory exchange of information pertaining to the ecologic characteristics of the polar regions, the populations living there, and the health problems of high latitudes and cold climates. The Symposium on Circumpolar Health Related Problems, sponsored by the Arctic Institute of North America and funded by the U.S. Public Health Service, was held in Fairbanks, Alaska, in July 1967 (see *Archives of Environmental Health,* vol. 17, October 1968). Medical and social scientists from eight nations sharing the arctic frontier participated, and a wide range of health problems and environmental stresses encountered in polar climates was considered in the symposium.

An international symposium on human biology and medicine in the Antarctic was held at the Scott Polar Research Institute, Cambridge, England, in September 1972. The proceedings have been published in *Polar Human Biology,* edited by O. G. Edholm and E. K. E. Gunderson, Heinemann Medical Books, London, 1973.

Although problems of adaptation to cold and social isolation are basically similar in the Arctic and the Antarctic, the Antarctic, because it has neither indigenous populations nor industrial or commercial enterprise and has much more severe environmental conditions, presents different types of physiological and behavioral problems for investigation from those encountered in the Arctic, where endemic and chronic diseases are of primary concern.

In spite of logistic and other difficulties, a number of investigators have concluded that the Antarctic offers many advantages as a laboratory for human biology. No other situation exists where groups of healthy individuals are isolated for several months and are available for physiological examinations on a regular basis.

PERSONNEL CHARACTERISTICS

Inhabitants of Antarctica typically are temperate zone dwellers temporarily transported to the antarctic environment for periods of 2 or 3 months (summer) or 1–2 years, after which they return to an urban environment in a temperate zone. The Antarctic thus represents a natural laboratory for the study of psychobiological adaptations of urban dwellers from temperate zones.

Men who winter in the Antarctic are not typical or representative of the general population. Having undergone rigorous qualifying medical examinations, they are highly selected in terms of physical health. Usually, they are highly motivated volunteers, and they have specialized educational and occupational backgrounds. Antarcticans tend to be youthful; participants range in age from 20 to 45, but most are under 35. There has been an increasing tendency for scientific field work to be done by graduate students or technicians rather than senior scientists. This trend has been noted at both U.S. and Soviet stations and is probably typical of other programs as well.

The U.S. program has differed from the programs of other nations in that military personnel were responsible for logistic support and station operation. U.S. Navy personnel provided transport, construction, maintenance, and vital service functions such as medicine, communications, and food preparation. The United States is moving toward an all-civilian program, and U.S. Navy involvement is now being phased out. Almost all other nations engaged in antarctic activities have civilians performing logistic support. In the New Zealand and Australian programs, and perhaps in others as well, a specialized corps of men experienced in polar work has developed over the years, and a few 'professional isolates' apparently spend much of their adult lives in the Arctic or the Antarctic [*Taylor,* 1969]. The typical member of an antarctic wintering party, however, is likely to be totally inexperienced in enduring cold weather and isolation.

The primary consideration in selection of antarctic personnel is occupational competence. Antarctic groups are generally composed of a variety of occupational specialists who may differ widely in cultural background and personality characteristics. Although the frontierlike existence lends itself to a friendly and informal atmosphere, the close quarters and the sameness of the physical and social environments tend to highlight personal and cultural idiosyncracies during the long winter. Morale and work efficiency are affected to some degree by the interpersonal frictions that inevitably arise. Thus psychosocial factors become important, and it is here that the greatest complexities exist and that knowledge is the least adequate.

The dual command arrangement at U.S. stations with two leaders, one military and one scientific, and two organizational subgroups of different composition has been a potential source of conflict and disruption. It would appear that, whenever a division of labor exists between scientists and support personnel, some degree of threat to cohesiveness will exist. Australian stations make a concerted effort to minimize status differences and factional loyalties.

Tolerance of arduous or relatively monotonous work would appear to be more likely when the experience is seen as being instrumental in achieving a worthwhile goal. The excellent performance of most volunteers for antarctic service suggests that the scientific purposes of the expeditions are generally accepted as being important.

ANTARCTIC ENVIRONMENT

There are actually many different antarctic environments. The widely dispersed stations on and around the continent exhibit a variety of climatic and topographic characteristics. To evaluate physiological or psychological observations and make comparisons from one site to another, it is essential to take into account the ambient conditions. At any station there are two aspects of the physical environment that need to be considered: the interior, or sheltered, spaces and the surrounding outdoor areas. Some of the same environmental variables need to be measured in both indoor and outdoor areas, and some parameters are unique to the indoors or the outdoors. An important characteristic of the interior spaces is the degree of confinement, that is, the volume of space available per individual. The constraints on movement, access, visibility, and so on imposed by the natural terrain surrounding the station may significantly affect station operations and individual living patterns. The degree of isolation (the distance from the nearest population centers and the time until relief in emergencies) and the duration of such isolation tend to compound the effects of other stresses.

Local climatic conditions vary with latitude, air movement (wind), and elevation. Coastal stations, located mostly between 60° and 70°S, have milder climates than stations on the interior ice cap unless extremely strong winds prevail at the coastal stations. Only the Americans and the Russians maintain permanent stations in the interior, where elevation, intense cold, and high winds combine to produce the most hostile environments inhabited by man.

The hazards and discomforts of antarctic living have been greatly reduced as protective clothing, shelter, food, and other life support items have improved. Everything needed for life except water (from snow) must be transported by sea or air to the station sites. Obviously, the quality of life support depends on the resources and the technology available. Safety and survival depend on preventive maintenance, preparation for emergencies, and constant vigilance. Fire is the greatest danger, and great care must be taken in the design of stations to limit damage to a small area if fire occurs.

Circulation of heat, ventilation, provision of water, and waste disposal have been significant problems at all stations, but the basic principles for managing these problems are well known. The methods used to meet these life support requirements differ among stations and can affect physiological status and behavior in many ways, particularly when the problems persist over long periods of time.

FUTURE RESEARCH

Problem areas in human biology and psychology that could be investigated in polar regions in the future with great benefit to medical and behavioral science were outlined in *Polar Research: A Survey* [*Gould,* 1970].

A complete inventory of worthwhile research problems related to human adaptation in polar environments would not be practical here. Many areas of interest and importance are presented in some detail in the following chapters, and the authors generally emphasize findings that merit further investigation. It seems likely that the research programs described in this volume will continue along the lines indicated but will gradually expand in scope to encompass a broad spectrum of issues in human biology, human ecology, and small social systems.

The goals of medicine in the Antarctic need to be oriented toward prevention of illness or disability and optimization of health rather than toward the traditional concern with curing disease. In this respect, antarctic medicine resembles aerospace and undersea medicine and is related to the emerging specialties of combat and sports medicine. Knowledge acquired in the Antarctic may have application in other unusual or restricted environments such as lunar colonies, orbiting laboratories, and experimental undersea stations. The importance of thermoregulatory, cardiovascular, and pulmonary mechanisms, as well as problems of isolation, confinement, and monotony, is common to a number of extreme operational settings. The deficiencies in biomedical and psychological knowledge needed for realistic planning of long-duration space missions have been stressed and elaborated on in the recent report of the *Space Science and Technology Panel* [1969].

Basic health and morbidity data have not yet been compiled for the antarctic environment. The relative risks of different types of illness and injury could be determined from routine medical records and epidemiological analyses involving all antarctic stations. International cooperation and interchange should be possible in medical research as it is in other fields, for example, in meteorology. Systematic collection of information on accidents and injuries in Antarctica has been recommended by the Scientific Committee on Antarctic Research ad hoc Subcommittee on Human Physiology.

Living in the Antarctic is certainly an unusual experience and contrasts sharply in many ways with urban life in temperate zones, but the physiological and psychological effects of spending a year or two in the Antarctic are still largely unknown. The investigation of long-term adaptive processes requires extended periods of observation beginning with a control period before the subjects live in the Antarctic and ending with follow-up observations after the subjects return home. This type of longitudinal research design is costly but essential for systematic investigations of human adaptation in Antarctica.

REFERENCES

Byrd, R. E., *Alone,* Putnam, New York, 1938.

Edholm, O. G., Medical research by the British Antarctic Survey, *Polar Rec., 12,* 575–582, 1965.

Edholm, O. G., and H. E. Lewis, Terrestrial animals in cold: Man in polar regions, in *Handbook of Physiology,* edited by D. B. Dill, E. F. Adolph, and C. G. Wilbur, pp. 435–446, American Physiological Society, Washington, D. C., 1964.

Gould, L. M., *Polar Research: A Survey,* Committee on Polar Research, National Academy of Sciences, Washington, D. C., 1970.

Silverberg, R. (Ed.), *Antarctic Conquest: The Great Explorers in Their Own Words,* Bobbs-Merrill, New York, 1965.

Space Science and Technology Panel, President's Science Advisory Committee, *The Biomedical Foundations of Manned Space Flight,* National Academy of Sciences, Washington, D. C., 1969.

Taylor, A. J. W., Professional isolates in New Zealand's Antarctic Research Programme, *Int. Rev. Appl. Psychol., 18*(2), 135–138, 1969.

Wilson, O., Human adaptation to life in Antarctica, in *Biogeography and Ecology in Antarctica, Monogr. Biol.,* edited by J. van Mieghem, P. van Oye, and J. Schell, W. Junk, The Hague, Netherlands, 1965.

PHYSIOLOGICAL RESEARCH AT BRITISH ANTARCTIC SURVEY STATIONS

O. G. Edholm

Division of Human Physiology, National Institute for Medical Research, London, England

Abstract. Physiological research at British Antarctic Survey (BAS) stations over the past decade is reviewed. Studies of food intake, energy expenditure, and weight loss or gain have indicated that approximately 3600 kcal/day is needed at base camp to maintain body weight, whereas approximately 5000 kcal/day is needed on sledging expeditions. Measurements of energy costs were made for many representative antarctic activities; indoor tasks involved low energy expenditure (generally less than 5.0 kcal/min), whereas outdoor activities such as skiing or digging snow involved high energy expenditure (approximately 9.0–10.5 kcal/min). Although antarctic diets included a high proportion of fat, blood cholesterol was not observed to increase. Ketonuria was common, however, probably owing to high energy expenditure and a calorie deficit. Studies of peripheral blood flow indicated decreased blood flow during winter months, providing evidence for local cold acclimatization in fingers and hands. It has been demonstrated that the Antarctic provides important opportunities to study a wide range of physiological effects of an extreme environment on man and to make regular observations that would be impractical elsewhere on groups of healthy young men over extended periods of time.

The British Antarctic Survey (BAS), known as the Falkland Islands Dependencies Survey (FIDS) until 1962, maintains a number of bases in the Antarctic, including a biological research center on Signy Island. The size of bases, as well as the number, varies from year to year; there are between 10 and 25 men at each base. BAS is a civilian organization that used to be under the Colonial Office but is now under the aegis of the Natural Environment Research Council. A number of doctors, as well as meteorologists, geologists, surveyors, etc., are appointed each year, and they spend a year, sometimes 2 years, at an antarctic base. Following discussion in 1956 an agreement was drawn up between the Medical Research Council (MRC) and the FIDS so that suitable medical officers could be trained in the MRC Division of Human Physiology to carry out a specific research program and so that, on returning from Antarctica, they could spend sufficient time in the Division to work up their results under guidance. This arrangement is still in force and has meant that a continuous program of guided research has been feasible. In recent years the subjects studied have broadened, and some of the doctors carry out their work from other laboratories while they retain links with the Division of Human Physiology. R. K. Macpherson, who was a senior member of the Division of Human Physiology, has established a similar system in Australia with the Australian National Antarctic Research Expedition.

Work carried out in British bases up to 1963 has been reviewed [*Edholm*, 1964, 1965; *Edholm and Lewis*, 1964], and only occasional reference will be made to studies covered by these reviews.

The position of the present British bases is shown in Figure 1. A new station has been opened in South Georgia, where the main emphasis is on biological research. The pattern of activity varies considerably among the different bases; some are essentially static with little possibility of travel, such as the base at Adelaide Island; others, such as Stonington Island, act as a base for traveling parties, and the number of people at base fluctuates considerably. A full account of the present British bases is given in the *BAS Handbook* (private circulation, 1970), reproduced in the appendix.

The objectives of the research program are to study the effects of the antarctic environment on man and to exploit a situation in which regular observations can be made on a group of healthy young men over a period of a year or more. In many cases the two objectives overlap. A primary physiological interest is to determine whether adaptation or acclimatization to low environmental temperatures can be demonstrated in man. It is therefore important to determine as accurately as is possible the conditions under which men live in the Antarctic, including their food intake and energy expenditure and hence their energy balance. The actual environmental conditions to which they are exposed must be assessed, and this assessment should include studies of clothing and shelter, i.e., the conditions at base and in the field. The effects that have been examined involve

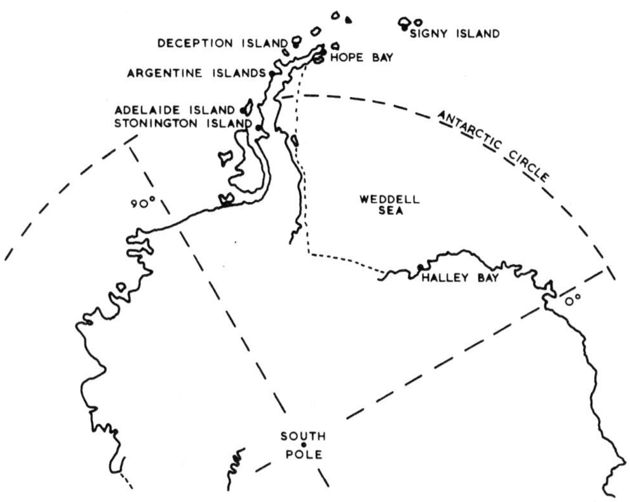

Fig. 1. Position of the bases maintained by the BAS at which physiological observations were made.

body and surface temperatures, peripheral blood flow, weight changes, and alterations in the thickness of subcutaneous fat, blood lipid levels, excretion of catecholamines, and sebaceous gland secretion. The environmental factors also include the daily pattern of light and darkness and the possible physiological effects of the long antarctic night in the winter compared with the effects of the continuous daylight of the antarctic summer. Another environmental factor is isolation, particularly in a vertually bacteriologically sterile atmosphere, as well as the effects that this may have on the incidence of infection and the levels of immunity. Since all the studies to be described have been carried out on members of the BAS, some of the physical characteristics of the population of the bases will be described first.

CHARACTERISTICS OF THE PERSONNEL AT BRITISH ANTARCTIC BASES

All the members of the BAS working in the Antarctic are volunteers. Although on some occasions members of the armed forces have been seconded to serve in the Antarctic, in recent years there have been scarcely any, and those few also have been volunteers. The majority spend 2 years in the Antarctic, although medical officers and some other specialists usually stay for 1 year only. They travel to and from the Antarctic by ship and are therefore usually absent from the United Kingdom either for 17–19 months for those who stay for a year or 29–31 months for the 2-year men.

The age range is from 19 to 40, but the great majority are between 22 and 29, and the average age (on arrival) is 25.5 years. Body weight ranges from 65 to 124 kg, the mode being 70 kg, and the mean weight is 74 kg. The height of the subjects varies from 162 to 189 cm, the mean being 179 cm. According to *Montegriffo* [1968] the average man in the United Kingdom aged 20–24 is 176.6 cm high and weighs 70 kg, so that the antarctic population is rather heavier and taller than the United Kingdom population, probably in part owing to the socioeconomic background of the antarctic group. They include doctors as well as geologists, surveyors, biologists, meteorologists, physicists, glaciologists, diesel mechanics, radio operators, cooks, and general assistants (Figure 2).

FOOD INTAKE

In the review by *Edholm and Lewis* [1964] there is a table giving the results of food intake measurements made at a number of antarctic bases. There is a considerable spread. Intake apparently averaged 4000 kcal daily, probably an overestimate. *Easty* [1967] carried out what is so far the most comprehensive survey at any antarctic base. He studied 25 men at Halley Bay at 6-week intervals, measuring total food intake over 24 hours. In all, 200 individual measurements were made. The mean daily intake was 3600 kcal, this being the mean value for the 200 measurements. The mean daily value for individuals ranged from 2670 kcal to 5260 kcal. Five of the men weighed between 85 and 100.5 kg, and their intake averaged 4500 kcal/day; the remaining 20 had an intake of 3375 kcal/day. There was also a large variation in an individual's intake throughout the year, and there was a significant seasonal variation. During the first 3 relatively warm 'summer' months (February, March, and April), daily intake averaged 3850 kcal. In the winter (May, June, July, and August), daily intake fell to 3360 kcal and rose again in the spring (September, October, November, December, and January) to 3660 kcal. This seasonal effect was exhibited by all subjects whatever their occupation. All the dietary intakes were measured at the base camp; although members of the expedition were engaged in sledging trips, these were of relatively short duration of 2–4 weeks, and these periods were not included in the survey. Details of the mean food intakes are given in Table 1.

At Signy Island (where most people are engaged in biological research) there was a mess survey of 12 men in May, August, and November (J. Brotherhood, unpublished data, 1969). Intake appeared to be high, 3900, 3600, and 3990 kcal being the average daily intake on each occasion. It is not uncommon for intake to be overestimated by using a mess survey technique. At Halley Bay, individual intakes were measured by J. Brother-

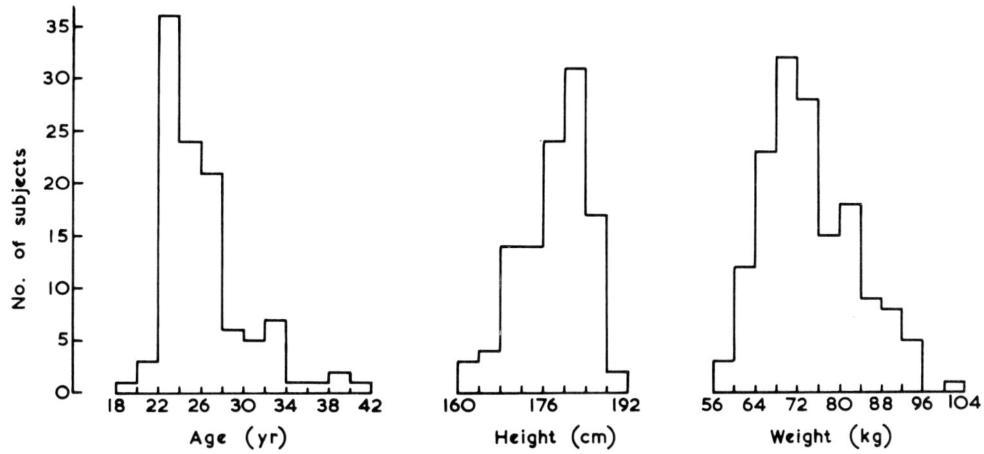

Fig. 2. Histograms illustrating the age, height, and weight distributions of the subjects studied in the Antarctic.

hood, but the results are not available. *Muir* [1969] recorded the food intake of five subjects at Stonington Island and obtained a mean value of 3700 kcal/day. *Lloyd* [1969] found that, on a diet of 3500 kcal/day, nine subjects during the winter at Halley Bay were in balance during two periods of study, each lasting 5 days.

At a British base camp there are no restrictions on food intake, and there is a wide variety of choice, so that food can be considered ad libitum. On sledging journeys the situation is different; with a sledging ration that has to be kept to a minimum weight and yet has to provide for a high-calorie intake, the fat content is considerable. Although there is some small variation from year to year, the BAS sledging ration provides approximately 4300 kcal/day, 55–57% of the calories being supplied by fat. Track vehicles are used for many journeys undertaken by members of the BAS, but dog teams are still employed, and occasionally man hauling is used. In the very rough terrain being surveyed, dog teams are considered not only more efficient than track vehicles but also safer in heavily crevassed country. Energy expenditure (considered in more detail below) can be very high, for example, in rough country where the sledges have to be man-handled and on occasions unloaded and reloaded. On the other hand, dog sledging can be very easy in travel over smooth snow, and the variability of energy expenditure is due also to the number of days when traveling is impossible because of climatic conditions, either blizzard or whiteout. Since many journeys are undertaken primarily to survey and map the country, the surveyor's needs, including good visibility, are dominant, and thus time spent in camp may be prolonged if the conditions are not satisfactory for the surveyor. For these reasons, studies of food intake during surveying have often proved difficult to assess, and thus a number of observations have been made on sledging journeys carried out primarily for physiological purposes. Since it has frequently been observed that men returning from a sledging journey have lost weight, *Orr* [1965] carried out a number of sledging journeys on which he varied the ration to determine the energy cost of sledging by feeding a ration on which the subjects did not lose weight.

The original sledging ration used by Orr provided no more than 4100 kcal daily, and men on this ration lost weight, which was rapidly regained on return to base. Similar changes have been observed by others [*Wilson,* 1960; *Massey,* 1956] and attributed to a relative dehydration during the journey with rehydration at base. *Wyatt* [1963], who also recorded substantial weight losses after sledging that were reversed within 2–3 days after the end of the journey, considered that dehydration cannot be the sole or even the main cause. Orr also concluded that fluid balance changes are an inadequate explanation. He therefore undertook several journeys during which he fed subjects on the standard ration for the first part of the journey and then measured the food intake on an unrestricted diet while they continued the journey. Although it is virtually impossible to standardize sledging during prolonged travel, on these journeys Orr succeeded in maintaining an almost constant mileage with a minimum of lie-up days. The results show that the weight loss during sledging was rapidly regained during the first few days at base but that water intake and urine output were not greatly different at base from those observed during sledging; thus Orr concluded that neither weight loss nor weight gain can be explained as being due to dehydration and subsequent rehydration.

Weights were recorded during the sledging journeys but only with an accuracy of approximately ±250 grams.

TABLE 1. Mean Values for Food Intake for 25 Subjects at Halley Bay

Intake, kcal	February/March	April	May	July	August	September	October/November	December
Total	3824 ± 836	3863 ± 1104	3434 ± 813	3346 ± 844	3301 ± 934	3729 ± 1067	3527 ± 892	3733 ± 821
Protein	528 ± 156	501 ± 187	392 ± 105	362 ± 101	366 ± 99	367 ± 99	389 ± 124	413 ± 413
Fat	1417 ± 448	1475 ± 614	1307 ± 429	1207 ± 337	1381 ± 503	1464 ± 502	1343 ± 517	1560 ± 400
CHO	1774 ± 406	1759 ± 500	1633 ± 364	1699 ± 506	1494 ± 479	1734 ± 489	1679 ± 449	1623 ± 370

Data are based on work by *Easty* [1967].

On the first journey, three men ate 4500 kcal daily, but, on a journey lasting 33 days and covering 620 km (385 miles), they lost, on the average, 2.5 kg; this weight loss was quickly regained when they had an unrestricted diet. On a second journey, three men covered 400 km (252 miles) in 70 days, and travel was much less arduous on this journey than on the first journey. Food intake was approximately 5000 kcal/day, and under the conditions of this journey there was a slight weight gain. Two of these men some 3 months later traveled 640 km (400 miles) in 19 days. This journey represented hard, continuous work. In good climatic conditions and on a standard ration of 3900 kcal/day, one man lost 5 kg by the sixteenth day, and the second lost 1.25 kg. They then ate an unrestricted diet while they continued the journey and in 3 days consumed approximately 9000 kcal/day. On this diet, weight increased toward the initial level.

Another journey was undertaken by two men who ate 4100 kcal/day for 13 days. Weight loss by the thirteenth day averaged 2 kg for the two men; then on unlimited rations they ate between 8000 and 9000 kcal daily and regained weight while they were still sledging.

On the next journey, three men used a daily ration supplying 5600 kcal/day and remained more or less in balance, although they were sledging hard and covering over 32 km (20 miles) a day. Orr has concluded that a sledging ration should provide between 5000 and 5500 kcal daily.

As was pointed out above, measurements of food intake during sledging in general show that all the usual rations provided are consumed but that there is a fall in body weight. Journeys of long duration have been undertaken successfully with rations yielding 4100–4300 kcal daily, probably because average energy expenditure has been reduced by the number of days when no sledging was possible.

ENERGY EXPENDITURE

The interpretation of the observations made on food intake both at base and during sledging requires, inter alia, information about energy expenditure. Although it can reasonably be assumed that, if body weight does not change over a period of weeks or months, food intake and energy expenditure have balanced, body weight changes are frequently observed, and thus estimates of energy expenditures are needed.

The methods used include time and motion studies, or, better named, timed activity studies, and measurements of oxygen consumption by different techniques. Timed activity studies, as the term implies, involve the observation and timing of all activities by the subject throughout 24 hours and usually include noting environmental conditions as well. The information obtained is expressed as a list of activities with the number of minutes occupied by each activity, the time spent indoors and outdoors, and the temperature recorded at the time of observation. From such records, energy expenditure can be computed from published measurements of the energy cost of each task, as listed by *Durnin and Passmore* [1968]. Since there is usually considerable individual variation in the energy expended in particular activities, the calculation of total energy expenditure can be in error for the individual, but estimates made for a group can be quite reliable. Accuracy is greatly improved if some measurements of oxygen consumption during a standard activity are made, since, in general, if an individual has a higher or lower than average oxygen consumption in one task, he is likely to have the same relationship to the average for other tasks.

Direct measurement has been made by using Douglas bags, the integrating motor pneumotachograph (IMP) [*Wolff,* 1958], the Max-Planck respirometer, and the Wright flowmeter. There are thermal limitations to be considered: ordinary mouthpieces conventionally used with Douglas bags freeze up at moderately cold temperatures. Face masks used with the IMP and the Wright flowmeter are liable to cause frostbite in cold weather on the areas compressed by the rim of the mask. However, with suitable insulation of the equipment, measurements have been made successfully out of doors at temperatures of -25°C with air speed up to 16 km/hr (10 mi/hr).

Norman [1965] studied four subjects at Halley Bay, making his observations for a 24-hour period once a month for each subject, using a specially designed diary card to record activity, clothing, and comfort sensations of the subjects. He also measured the subclothing temperature and during sledging recorded food and water intake.

The energy expenditure calculated from the activity records was, on the average, 3400 kcal daily for the four men over the whole year. In the summer months (October–April) the expenditure rose to 3770 kcal/day, and in the dark winter months (May–September) it fell, on the average, to 3120 kcal. Norman also calculated that energy expenditure during sledging (man hauling) averaged 5050 kcal/day. On a food intake averaging 4430 kcal daily there was a mean weight loss of 1.25 kg after 5 days of sledging. *Wyatt* [1963] calculated the energy expenditure during a number of sledging expeditions; on 22 out of 23 man journeys there was a loss of body weight averaging 3.8 ± 2.3 kg. The ration provided 4100 kcal daily, but on some of the longer journeys the ration was cut, and then the daily food intake averaged only 3740 kcal. However, the mean daily energy expenditure was affected by the number of lie-up days, which amounted to 25–40% of all days, on different journeys and was estimated to be between 4000 and 4500 kcal daily.

Apart from the measurements made by J. M. Adam and by A. Rogers, already reported [*Edholm and Lewis,* 1964], no detailed measurements of oxygen consumption have yet been published. Recently, *Brotherhood* [1973] completed a 2-year study in Antarctica, the first year being at Signy Island and the second being at Halley Bay. During this time he made a large number of measurements of the energy cost of various tasks representative of activities in the Antarctic (Tables 2 and 3). Of particular interest is the energy cost of walking, which can vary twofold (5.98–11.25 kcal/min) according to surface conditions and wind speed. These figures indicate the difficulty of giving average figures, particularly for activities out of doors. However, when calorie cost is plotted against speed of walking at Halley Bay, much of the scatter is accounted for. Also, comparison with figures for the cost of walking at different speeds, as illustrated by *Durnin and Passmore* [1968], shows that the energy cost of walking is nearly twice as high in the Antarctic as it is in temperate regions. The subjects walked on a level 1.6-km (1-mile) course, measurements being made during the whole period of the walk in both directions (Figure 3). The surface consisted of moderately firm snow on ice, footprints being from ½–2 cm deep, and these conditions were more variable and more difficult at Signy Island than at Halley Bay; snow with a weak crust tended to be deeper.

At walking speeds, which varied only between 3½ and 4 km/hr, energy costs ranged from 5 to 12 kcal/min. The average figure of 9.1 kcal for dog sledging (Table 2) is rather lower than previous estimates, but at Halley Bay the measurements were made during sledging on fairly level ground and did not involve unloading and reloading the sledge or man hauling over rough ice hummocks. Many of the activities associated with sledging, including ski walking, preparing camp, cutting up seal meat, and feeding dogs, involve high rates of energy expenditure and give support to the high estimates of food requirements for sledging journeys quoted above.

Ski walking involved moderate to high levels of energy expenditure ranging from 7.5 to 14.2 kcal/min and averaging 10.6 kcal. During skiing lasting 6–8 hours, average expenditures of 9.0 kcal/min were maintained.

Activities at base were very variable among individuals, but all had to carry out a number of common tasks, some of which involved high energy expenditure, such as digging snow and filling water tanks (10.5 kcal/min),

TABLE 2. Energy Expenditure in Kilocalories per Minute

	Signy Island		Halley Bay	
	Expenditure	Range	Expenditure	Range
	Indoor Activities			
Domestic				
Washing up and serving	3.2 (8)	2.8–4.2	4.1 (13)	2.4–5.8
Sweeping and tidying	5.1 (8)	4.2–5.7	6.4 (7)	5.5–8.3
Washing (machine)			3.55 (2)	3.5–3.6
Sewing			1.60 (1)	
Workshop				
Bench work	2.1 (1)		3.3 (11)	2.3–4.1
Cutting metal	3.6 (1)			
Using circular saw			4.3 (2)	3.7–4.9
General carpentry			5.0 (11)	4.5–6.0
Servicing tractor wheels			4.3 (3)	3.3–5.0
Sledge repairs			3.3 (16)	2.2–4.6
Laboratory work				
Sitting	1.95 (17)	1.45–2.5	1.9 (7)	1.5–2.1
Sitting, using typewriter or calculator	2.4 (2)	2.2–2.6	2.5 (3)	2.3–2.75
Standing	2.6 (9)	2.1–3.2		
Radio operator			2.0 (2)	1.95–2.05
	Outdoor Activities			
Working in snow				
Filling tanks with snow	10.4 (15)	8.1–12.4	9.0 (3)	8.6–9.5
Digging snow	9.7 (8)	7.2–10.75	10.3 (3)	9.0–11.7
Digging out dumps			8.7 (4)	7.2–10.0
Moving stores			9.7 (6)	7.6–14.0
Loading fuel drums			8.0 (5)	6.0–10.6
Carrying rubbish sacks	10.8 (3)	9.3–12.3		
Fieldwork				
Meteorological observation	6.5 (7)	5.2–7.2		
Botany	7.2 (9)	5.6–8.1		
Marine biology on sea ice				
Cutting holes in ice	6.4 (2)	6.2–6.6		
Fishing (hand line)	5.4 (11)	3.75–7.5		
Hauling and laying trawls	9.1 (8)	5.1–11.5		
Trawling and dredging	9.2 (13)	6.5–11.0		
Marine biology in boat				
Sitting in dinghy	2.75 (9)	1.7–3.5		
Fishing (hand line)	2.8 (2)	2.0–3.7		
Trawling and hauling	5.3 (4)	4.2–7.0		

Measurements were made by *Brotherhood* [1973]. The number of observations is given in parentheses.

digging out dumps and moving stores (8.5–10.5 kcal/min), and cutting up frozen seals and feeding dogs (8.0–10.0 kcal/min). Indoor activities apart from building work required much lower energy expenditure, the highest being 6.4 kcal/min for sweeping and washing floors. Other indoor activities ranged from 1.5 to 5.0 kcal/min.

At Signy Island the level of activity was relatively constant throughout the year, since some outside work was always feasible. The overall daily energy expenditure was of the order of 3600 kcal for the whole year. There was, nevertheless, a considerable spread among individuals.

The situation at Halley Bay was different, and the activity level was substantially higher in the summer than in the winter, when outside work was difficult and restricted.

BODY WEIGHT CHANGES

A number of observations in which seasonal weight changes have been observed on members of polar expeditions have already been reported [*Lewis et al.*, 1960; *Wilson*, 1960]. These changes consist of an increase in body weight during the winter and a fall in the summer, and corresponding changes in skinfold thickness, reflecting alterations in the amount of subcutaneous fat, also have been observed. The commonly accepted explana-

TABLE 3. Energy Expenditure for Skiing, Sledging, and Camping in Kilocalories per Minute

	Signy Island		Halley Bay		Stonington Island
	Expenditure	Range	Expenditure	Range	Expenditure
Ski walking					
Sea ice	11.4 (12)	8.0–14.3			
Land	11.6 (5)	9.8–13.3	8.9 (10)	7.5–10.5	8.0
Skijoring			6.4 (2)	6.0–6.75	4.8
Sledging					
Driving motorized sledge (Skidoo)	3.1 (22)	2.4–4.2			
Sledge hauling	11.8 (5)	8.3–13.1			15.2 (uphill)
Dog sledging			9.1 (5)	8.0–10.5	
Sitting on sledge			4.5 (2)	4.0–5.0	
Running and sitting with sledge					7.1
Walking with sledge					6.9–8.6
Camping					
'Outside' man			8.5 (3)	8.0–9.5	5.6–6.6
'Inside' man			5.0 (1)		
Feeding dogs					
Chopping frozen seal			8.0 (3)	5.7–10.0	9.5
Sawing (crosscut) frozen seal			9.7 (6)	8.5–11.4	
Sawing (power saw) frozen seal			6.3 (2)	6.1–6.5	
Feeding dogs			8.6 (4)	7.3–9.9	

Measurements at Signy Island and Halley Bay were made by *Brotherhood* [1973], and those at Stonington Island were made by A. L. Muir (unpublished data, 1969). The number of observations is given in parentheses.

tion is that the usual seasonal change of energy expenditure is responsible, although it has been suggested that the increase in fat thickness during the winter represents a biological adaptation to cold. As was pointed out by *Edholm and Lewis* [1964], the observed changes are small, and the additional thermal insulation is trivial. However, the existence of the phenomenon has been called in question by *Orr* [1965], who made observations at Hope Bay over a period of 2 years. Although changes of weight were observed in most of the 20 subjects, these were irregular and were not related to season. Orr considered that the lack of a seasonal change was due to the very active sledging program carried out through the whole year, since Hope Bay is relatively far north (63°24′S).

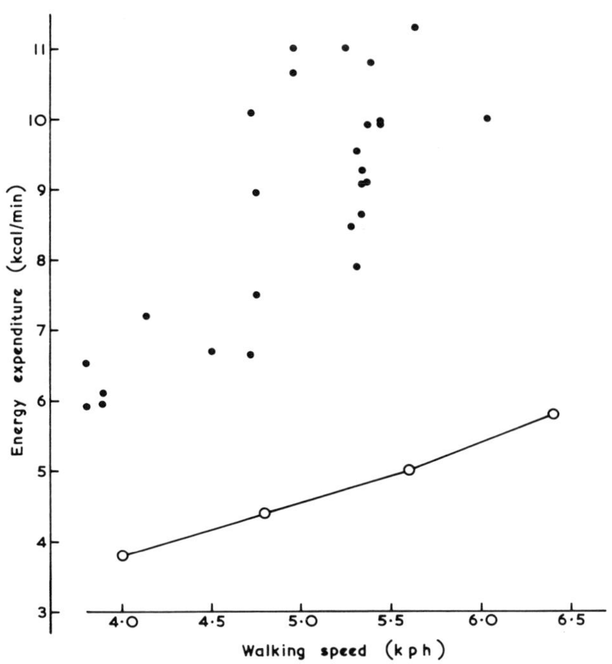

Fig. 3. Energy cost of walking on a firm, level surface at Halley Bay (filled circles). The average cost of walking at these speeds, determined in the United Kingdom [*Durnin and Passmore,* 1968], is also shown (open circles).

Easty [1967] recorded the body weight of 24 men at monthly intervals from February until the following January. The average body weight increased by 3 kg from February until August and then declined but only slightly. All subjects gained weight, but there were considerable individual differences, the maximum weight gain ranging from 1 to 6.75 kg, the average being 4.4 kg. There was no evident relationship between initial body size and weight gain. Fifteen of the 24 men reached their maximum weight in the period from August to October, but three subjects recorded their highest body weight in January before the relief ship arrived. There were only three men whose final weight in the Antarctic was not higher than their weight on arrival.

Skinfold thickness also was measured at four sites (subscapular, lateral arm, abdomen, and chest). The mean of the sum of these for all subjects shows changes similar to the alterations in body weight (Figure 4).

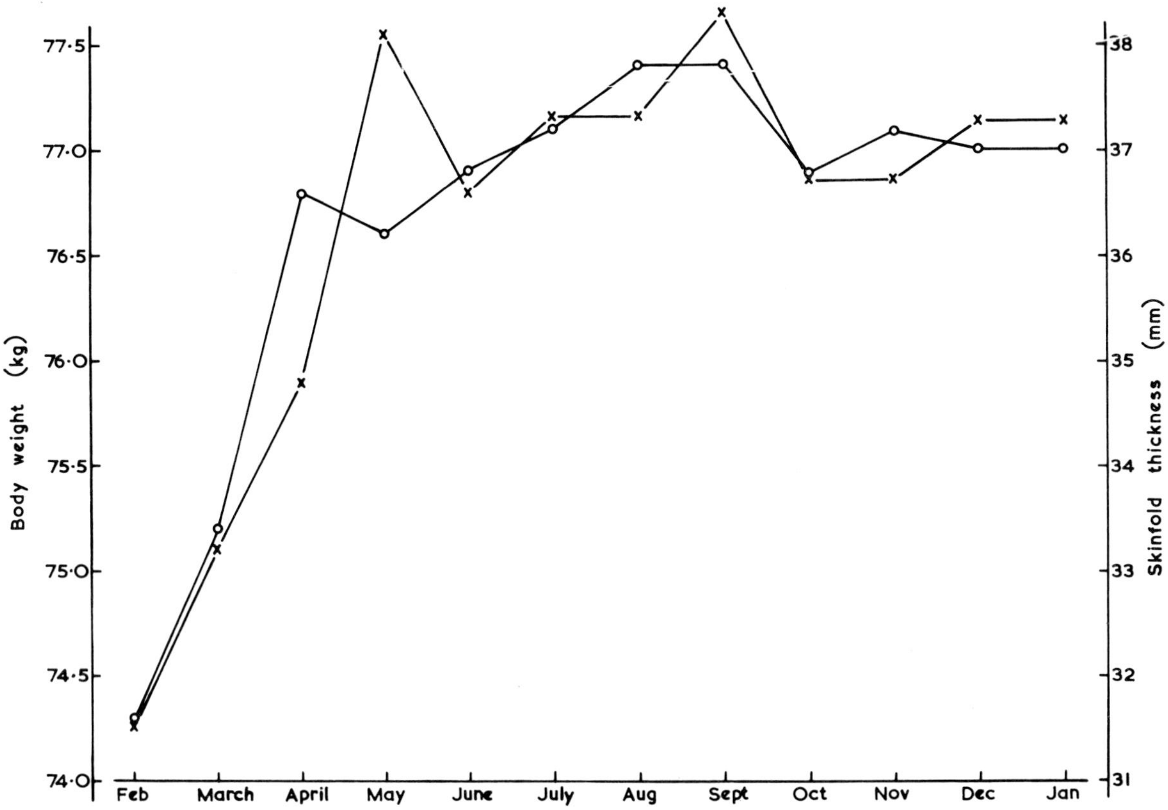

Fig. 4. Seasonal changes in body weight (open circles) and skinfold thickness (crosses) of 25 subjects at Halley Bay [*Easty*, 1967]. The skinfold thickness equals the sum of the fat thickness at subscapular, pectoral, lateral arm, and abdominal sites.

There was considerable seasonal variation in activity at Halley Bay; during the winter months it is not feasible to spend long periods out of doors. On the other hand, at Hope Bay, sledging activity continued all year, and *Orr* [1965] did not find evidence of seasonal changes, although most of his subjects had considerable fluctuations in body weight. Indeed, the lowest body weights were observed in July and August, and the highest were observed in September and October. However, it is difficult to draw firm conclusions from Orr's figures, since body weights at base could be recorded only at irregular intervals.

It is therefore of interest to compare records made by J. Brotherhood (personal communication, 1969) in successive years at Signy Island and at Halley Bay. The pattern of activity at Signy Island is fairly constant throughout the year, since it is situated outside the antarctic circle and outside work continues throughout the year. At Halley Bay there is clear evidence of a seasonal pattern, activity diminishing in the winter months. A seasonal curve is evident at both bases, and, since Brotherhood was able to record body weight under carefully controlled conditions at 2-week intervals on all subjects, considerable reliance can be placed on his records (Figure 5).

A. G. Davies [1969] studied seasonal changes in body weight and skinfold thickness in four subjects at Stonington Island. All four were engaged in long sledge journeys; these continued throughout the year, but the major journeys were in the spring and summer. One man spent 34 days sledging in the autumn (April–June), 10 in the winter (July–September), 54 in the spring (October–December), and 34 in the summer (January–March). Body weight and skinfold thickness variations were in general highly correlated; however, there was no evidence of a weight gain during the winter. In the spring there was a marked fall in weight and then a rise during the summer. Davies concluded that the changes that he observed were due to alterations in energy balance.

The differing curves of body weight can be compared with the detailed records described by G. M. Budd and obtained at the Australian bases. Budd's comment that 'weight increase in the winter occurs often enough to be suggestive but is by no means invariable' can be applied also to the changes observed at British bases [*Budd*, 1974, p. 47]. He draws attention also to the variations between different bases as well as the effects of sledging or traveling and the effect of different programs of work in successive years. It must, in spite of extensive investigation, be concluded that the existence of a true sea-

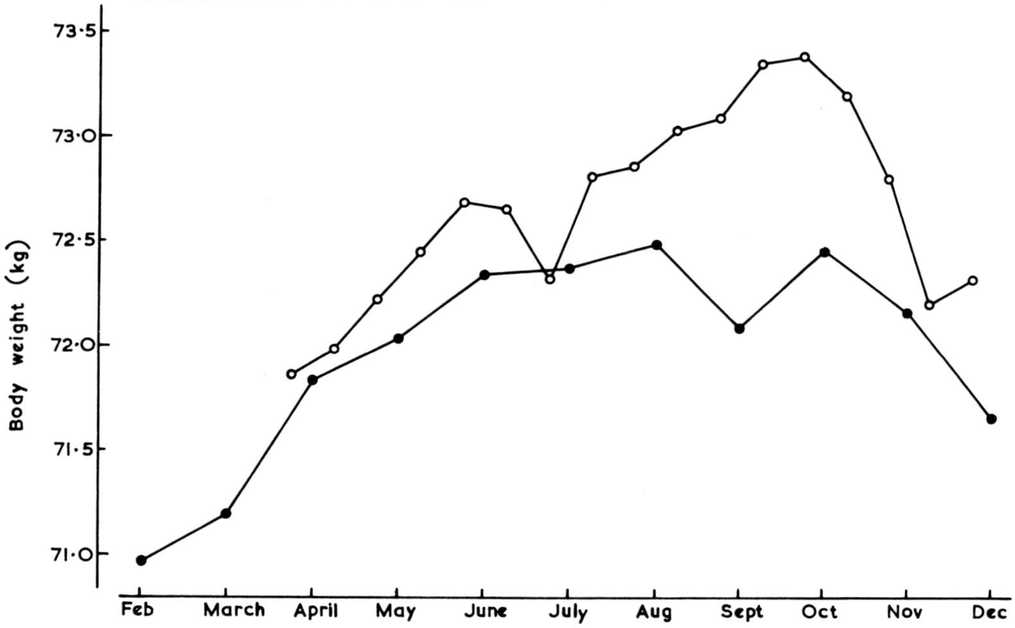

Fig. 5. Seasonal changes in body weight observed by J. Brotherhood (unpublished data, 1969) at Halley Bay for 14 subjects (open circles) and Signy Island for 11 subjects (filled circles).

sonal change in body weight and body fat is still an open question.

BLOOD LIPIDS AND FAT COMPOSITION

The relationships between environmental factors and blood lipid levels are of considerable clinical importance in view of the association described between cholesterol levels and coronary heart disease. It has been claimed that physical activity reduces the likelihood of coronary occlusion and also that activity reduces blood cholesterol levels. The increasing incidence of coronary heart disease and atherosclerosis has been considered to be due to the increasing fat content of the usual diet in the western world.

Some of the environmental factors can conveniently be studied in the Antarctic, where food intake, as has already been shown, is relatively high. Although the proportions of fat, protein, and carbohydrate in the diet consumed at base are similar to those found in the United Kingdom and the United States, the total quantity of fat consumed is relatively large. During sledging of course, when 53–57% of the calories are derived from fat, the quantities consumed are considerable. Answers have been sought to questions concerning the existence of seasonal variation in the various blood lipids, the effects of a cold environment, and the effects of sustained physical activity during sledging.

Blood cholesterol levels have been examined by *Antonis et al.* [1963, 1965] at Halley Bay and at the South African base, Sanae, and by *Hicks* [1967] at the Australian station, Wilkes. The mean levels, 199 mg/100 ml (24 subjects) at Halley Bay, 215 mg/100 ml (10 subjects) at Sanae, and approximately 250 mg/100 ml (23 subjects) at Wilkes, differed slightly. There was a slight downward trend, from 210 to 191 mg/100 ml, during the year at Halley Bay, and this was more noticeable in the first-year subjects (211–187 mg/100 ml) than in the eight second-year men (207–199 mg/100 ml). The values for the South African subjects were more variable, but the initial and final values were 217 and 205 mg/100 ml. Hicks concluded that at Wilkes station there was only slight variation; nevertheless, his figures also suggest an overall downward trend. It does not appear that the relatively large consumption of fat throughout the year has in any way increased blood cholesterol. Apart from the slight downward trend in total blood cholesterol over the whole year there was possibly a small increase during the winter months, but a more evident seasonal effect was shown by the changes in β and α cholesterol [*Antonis et al.,* 1965], particularly in the first-year subjects at Halley Bay. There was a definite and significant rise in β cholesterol during the winter and a concurrent decrease in α cholesterol (Figure 6). *Easty* [1963] also examined the effect of sledging journeys. Total cholesterol in 16 subjects fell from 206 mg before sledging to 191 mg immediately after sledging (the average length of journey was about 3 weeks). However, there were more marked effects on β cholesterol, which fell from 168 to 144 mg, and α cholesterol, which rose from 38 to 47 mg. These changes were statistically significant. In all cases,

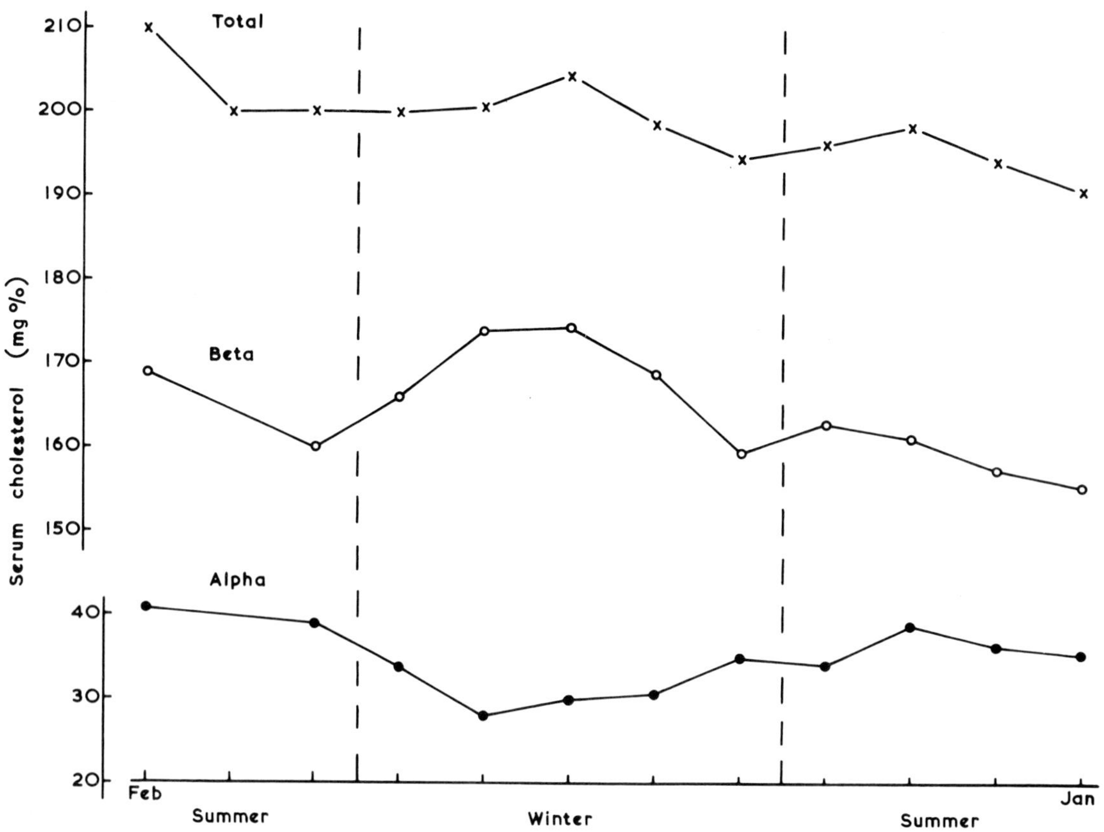

Fig. 6. Seasonal changes in total as well as α and β cholesterol (25 subjects) observed at Halley Bay [*Antonis et al.*, 1965].

levels returned to presledging values in a month. It is likely that the counterchanges in α and β cholesterol, together with the small rise in total cholesterol in the winter, were due to the relative inactivity at this time and that the sledging produced changes because of the increased physical activity. It may be concluded that living in a very cold environment does not influence cholesterol levels to any significant extent.

No seasonal changes were observed in phospholipids at either Halley Bay or Sanae. The mean levels were 229 mg/100 ml and 238 mg/100 ml, respectively, and the monthly variations were very small. There was more variation in the levels of triglycerides, which averaged 121 mg at Halley Bay and 115 mg at the South African base, but there was no seasonal trend. Thus, in spite of marked variations in physical activity and dietary intake, serum lipids in general were remarkably constant, the important exception being α and β cholesterol. There were striking individual variations in the levels of blood lipids, and such variation was most apparent in the triglycerides. In the 24 subjects at Halley Bay, 19 had a mean blood level of 91 mg/100 ml blood, and the remaining five had average levels of 231 mg/100 ml blood. In this group of five subjects, triglyceride levels varied greatly from month to month, whereas in the 19 subjects with low values these were fairly constant. The high-triglyceride group were significantly heavier, with thicker subcutaneous fat layers, faster resting pulse rates, and higher blood pressures than the low-triglyceride group. They had a higher food intake, but this was not statistically significant; blood cholesterol level was also higher but not significantly.

Easty et al. [1967] also obtained fat biopsies at regular intervals from 19 subjects. Animal work has shown substantial differences in fat composition that were consistent with adaptation to a cold environment, and there have been suggestions that similar differences may exist in man. Easty was interested in the possibility that changes might occur in fat composition during the course of a year spent in the Antarctic. The fat was analyzed to determine the proportions of the different fatty acids, and the results were compared for three seasons: summer (February–April), winter (May–September), and summer (October–January). The only fatty acid showing a significant (at the 5% level) seasonal change was oleic acid, which fell during the winter.

Small changes were observed between seasons in other fatty acids, but none were significant. However, comparison of the composition of fat in different groups showed that, in the Halley Bay subjects, linolenic acid was present (2.2%), whereas only traces or very small quantities have been found in the subjects of other surveys. No seasonal change in linolenic acid was observed, and it was present in the first samples obtained, i.e., within a month of arrival in the Antarctic. In other respects the results from Halley Bay were similar to those obtained in surveys in the United Kingdom and the United States, so that, with the doubtful exception of oleic acid, there was no evidence of change in the composition of subcutaneous fat that could be attributed to effects of the environment.

KETOSIS

The normal metabolism of fatty acids leads to the formation of small quantities of ketone bodies, of which the most important is acetoacetic acid. However, on a high-fat diet or during starvation or severe physical activity the production of ketone bodies is increased and may be in sufficient quantity to disturb acid-base balance in the blood. This condition is known as ketosis, or acidosis, and, depending on the degree of disequilibrium of acid-base balance, can cause a variety of symptoms ranging from mild malaise to coma. During sledging, men have a high level of energy expenditure, are on a high-fat diet, and are frequently in calorie deficit, a combination of conditions that might be expected to produce a degree of ketosis.

Muir [1969] has reported the results of a preliminary experiment in which he measured the urinary ketone bodies in five men during winter sledging from Stonington Island (68°11′S, 67°00′W). Food intake was measured, and a detailed record of activities was kept, from which an estimate of energy expenditure could be made. In addition, Muir measured ventilation volumes and used his results to estimate the energy expenditure of various sledging journey tasks (Table 3).

The excretion rate of ketone bodies in the urine was significantly related to energy expenditure during sledging (r = 0.624, $p < 0.001$). However, on days of great activity the subjects were also in calorie deficit, a factor known to increase the formation of ketone bodies, and in addition were exposed to low temperatures. Muir concluded that the high-fat sledging ration was not a primary factor, since the excretion of ketone bodies was low during days when travel was impossible (lie-up days), although the same diet was consumed. On the coldest journey, ketone excretion rates were again low, so that on Muir's evidence it would not appear that cold exposure was an important factor in producing ketosis.

A more detailed study was made by *Lloyd* [1969] at Halley Bay. He measured the excretion of ketone bodies by nine subjects while they were living at base either on a high-fat diet (60% fat, or approximately 225 g/day) or a diet containing only 30% fat (about 113 g/day). Seven of these subjects were studied also during sledging, when for 5 days they had a diet containing 60% fat (about 285 g/day) and after an intervening period of 4 days were on an ad libitum diet with a low-fat ration (30% fat, or approximately 142 g/day). Energy expenditure was estimated from a timed activity study.

When all the results were pooled, diets being ignored, a significant relationship was found between ketone excretion rate and energy expenditure, Muir's findings thus being confirmed and extended.

Muir had pointed out that on days of high energy expenditure his subjects were in negative calorie balance. Lloyd was able to investigate this aspect in more detail; again pooling all his results, he showed that there was a highly significant correlation between a negative calorie balance and the excretion of ketones. There was no demonstrable correlation between ketonuria and the fat content of the diet, but it appeared that ketonuria was more common on days of great activity on a high-fat diet than on days of similar activity on a low-fat diet. There was also a significant association between cold exposure, assessed as time spent out of doors, and ketonuria, but this effect is slight when the factor of calorie deficit is excluded. The causes of ketosis in the Antarctic are multiple; the most important are high energy expenditure and calorie deficit, although exposure to cold is a further but minor factor. The fat content of the diet appears to be of little importance, although it may be a minor factor in calorie deficiency.

Both Muir and Lloyd have emphasized how relatively common ketonuria can be in the Antarctic. It may be responsible for some complaints of malaise, but the harmful effects are to a large extent ameliorated by the frequency of forced rest days during sledging. It has been shown in animal experiments that rats acclimatized to cold develop less ketosis than unacclimatized animals. Both Lloyd and Muir tried to find out if there was any evidence of acclimatization but were unable to get any clear results.

These findings on ketosis demonstrate the importance of maintaining calorie balance, especially during sledging. They are also of interest in connection with survival in polar regions and survival rations. *Rogers et al.* [1964, 1966] and *Rogers and Aksnes* [1969] have carried out a series of experiments in the Arctic on survival rations, including periods of 5 days without food. The severe ketosis that developed was accompanied by a diminution of extracellular fluid. Sodium was excreted in large amounts; it was found that its excretion could be prevented by giving 100 mEq/day in the form of sodium bicarbonate. The fluid loss was virtually abolished, and the clinical effect of starvation was diminished. Without the supplement of sodium bicarbonate, with the onset of diminished extracellular fluid most subjects had pronounced falls of blood pressure on standing. Preventing fluid loss also diminished this orthostatic intolerance. A survival ration was developed and tested satisfactorily with a calorie intake of 500–1000 kcal/day provided by sugar and a supplement of sodium bicarbonate. It seems possible that on sledging journeys with a significant deficit of calories there may be a loss of sodium chloride and a fall in extracellular fluid that could aggravate the ketosis.

CARBOHYDRATE METABOLISM

Recently, an interesting dietary experiment was carried out successfully at Halley Bay. Food in which sucrose was replaced by glucose was prepared for consumption at base. Sucrose intake was reduced from an average of 120 g/day to less than 5 g/day, approximately 100 g glucose being added to the diet, and the subjects stayed on this diet for 8 weeks. Glucose tolerance tests were carried out at weekly intervals, and a statistically significant difference was observed after 3 weeks of the glucose diet, blood glucose rising to a lower level on the sucrose-free diet than on the sucrose diet [*Fry,* 1971]. Blood lipids also were measured, and it appears that glycerides tended to fall at least initially on the sucrose-free diet. Glyceride level rose toward the initial level while the subjects were still on the sucrose-free diet, but then, on change to the normal diet, i.e., with sucrose instead of glucose, glycerides continued to rise for some 8 weeks. Cholesterol levels showed very little change [*Roberts,* 1971]. This study is another example of using the Antarctic as a convenient laboratory for carrying out an experiment that could be done in a temperate climate, but the practical difficulties there would be considerable.

PERIPHERAL CIRCULATION AND COLD EXPOSURE

Hampton [1969*a, b*] during 2 years spent at Hope Bay used a calorimeter technique for measuring hand blood flow to find out if there was any evidence of changes in the peripheral circulation as a result of local acclimatization to cold. *Massey* [1959] has already confirmed at Hope Bay Mackworth's original observation regarding the development of a resistance to cold in the fingers, as shown by a diminished loss of cutaneous sensation, in subjects whose hands were habitually exposed to cold. This has rightly been accepted widely as the clearest evidence of local acclimatization to cold that we possess. It would be tempting to believe that blood flow in the fingers might be correspondingly higher in the acclimatized than in the unacclimatized. In addition, many workers have examined the cold vasodilation response that follows immersion of the fingers in ice water and have found that there is a less complete initial vasoconstriction, a more rapid increase in blood flow, and apparently much less pain in subjects whose hands are habituated to cold (fishermen and fish filleters) than in other subjects.

Hampton was able to follow four subjects for 2 years in the Antarctic and a second group of four for 1 year to measure hand blood flow by using a calorimeter technique at frequent but irregular intervals. The observations were always made in the early morning when the subjects were lying comfortably in bed, since the subjects could be maintained in relatively similar environmental conditions most effectively in this way. The subjects spent a high proportion of their time sledging and so were considerably exposed to low temperatures, particularly in their hands. There was considerable variation in the levels of peripheral blood flow both among individuals and in the same subject from time to time. Nevertheless, clear evidence of a seasonal change was obtained: blood flow declined during the winter months, although the level at the end of the time spent in the Antarctic was similar to the level on arrival. Hampton has pointed out that these findings conflict with previous views about the effects of local acclimatization to cold in the fingers and hands, since it has been concluded that the blood flow in the hands would be increased, not diminished. However, Hampton's results can usefully be compared with *Elkington*'s [1968], described by *Budd* [1974]. It is of great interest that Elkington also obtained a decrease of finger blood flow in the colder period of the year, although the technique and procedure were quite different from those used by Hampton. Elkington examined finger blood flow before and during immersion in ice water; Hampton measured heat flow from the hand in water at 20° and 30°C. Both concluded that there is an increased vasoconstrictor tone during the antarctic winter, and this conclusion agrees well with Budd's own observations.

BLOOD PRESSURE

There have been reports that blood pressure can increase considerably in those who are exposed to cold for a long time. However, since these reports have been based on quite inadequate evidence, *Davies* [1967] carried out a survey using a novel form of blood pressure measurement by which frequent readings could be obtained with little disturbance of the subjects [*Davies and Lewis*, 1967]. His work was done at Adelaide Island, where sledging is not feasible, although there is a good deal of skiing, so that the degree of cold exposure was probably less there than it would have been at Hope Bay, for example. In general, there was no conclusive evidence of any seasonal effect nor of any marked effect of the antarctic environment when a comparison was made with blood pressure recorded in the same subject in the United Kingdom. There was a tendency for systolic pressure to be higher in the winter than in the summer, but the difference was not significant. The mean systolic pressure recorded throughout the year was some 3 mm Hg higher in the Antarctic than in the United Kingdom, but this difference was not significant. Diastolic pressure also was lower in the United Kingdom but was not significantly so. Pulse pressure appeared to be similar in the Antarctic and the United Kingdom.

Casual measurements have been made by a number of medical officers in the Antarctic, but there has been no evidence of an increase in blood pressure. Such negative evidence does not preclude the possibility of the development of hypertension as a result of the difficulties imposed by the antarctic environment, since these difficulties are not easy to assess or quantify. Some studies have been made, however, in which, inevitably, the word 'stress' has been used.

STRESS OF THE ANTARCTIC ENVIRONMENT

The use and abuse of the term stress have been discussed by *Edholm and Bacharach* [1965], but as far as possible this engineering term should be applied to the environment producing a resulting 'strain' in man. The components of the antarctic environment that might cause strain include low temperatures and strong winds, the dangers of traveling, and the problems associated with the isolation of a small group of young men. This isolation is dealt with in detail by *Natani and Shurley* [1974].

Among the studies made so far, *Simpson* [1959] assessed adrenocortical activity indirectly by measuring levels of circulating eosinophils and recording the subjects' actions and situation. He was able to show some relationship between the situation and changes in the eosinophil count, one situation involving the marked reactions of a leader who was not effective and was apparently anxious, and the evidently unpopular task of cooking.

C. Brown (personal communication, 1966) used flicker fusion frequency and card sorting to find out if there is any evidence of seasonal changes at Halley Bay. No significant changes were observed. More recently, *T. W. Davies* [1969] used a personality inventory, together with a regular recording of mood at intervals throughout the year, to classify the members of an antarctic base. He also estimated catecholamines by measuring the excretion of vanillyl mandelic acid (VMA) in the urine. He found a marked diurnal rhythm in the quantity of VMA in the urine, i.e., low levels at night that rose to a peak in the early afternoon. There was some evidence of a small increase in daytime excretion in the winter, but night levels remained constant. Davies grouped his subjects on their scores for extroversion and introversion but found little difference between them in excretion rates. Examining the relationship between mood and VMA in the urine, he found that moods indicating tranquility or happiness were not related but that moods of anger, fear, or boredom were in several but not all subjects associated with increased excretion of VMA. Since the number of subjects was only nine, Davies' results can be regarded only as being suggestive. Moods of anger, fear, or frustration were relatively rarely acknowledged, and boredom was the most commonly recorded 'negative' mood. There were suggestions that those scoring high in extroversion experienced more boredom than introverts, as might be expected, but there was also suggestive evidence that the extroverted when they were bored excreted more VMA than the introverts when they were bored. The evidence is really inconclusive, since introverts so seldom acknowledged a state of boredom. Nevertheless, it would be useful to repeat Davies' work on a larger scale in an attempt to link physiological and psychological studies.

As was mentioned above, there was a small overall increase in the excretion of VMA in the winter, but the mean value for VMA excretion in the Antarctic was similar to that in the United Kingdom in three subjects.

SLEEP

Lewis and Masterton [1957] reported that the average duration of sleep as recorded on diary cards was 8 hours a day over a period of 2 years in northern Greenland. They also showed that sleep patterns in periods of complete darkness were different from those in periods of complete light. In these seasons, sleep was not taken in one spell but was broken up into several shorter periods. There was also some evidence that the length of the day could exceed 24 hours. Several sleep studies using the same technique have been carried out in the Antarctic, and the average duration of sleep once again comes close to 8 hours. However, the disturbance of sleep patterns during the long antarctic winter and summer has not been observed. This undisturbed sleep may be due to the regularity of the regime at antarctic bases, although this does not prevent a considerable individual variation in the duration of sleep.

C. Booker (personal communication, 1960) at Halley Bay found that the number of hours of sleep per person varied from 6 to 9½ and that there was considerable constancy in any one individual but there were significant differences between individuals in their number of hours of sleep averaged over the whole year. Booker also recorded weight changes, and it was expected that weight gains would be greatest in the long-sleep subjects and least in the short-sleep subjects, but no such relationship was observed.

WATER BALANCE AND COLD DIURESIS

In the section on body weight changes, mention was made of the fall in body weight during sledging and the rapid regain on return to base. Although it is now clear that such changes reflect calorie deficit followed by a period of positive calorie balance, nevertheless, a degree of dehydration during sledging, followed by increased fluid intake at base, also has been postulated as the cause. It has been suggested also that urine volume during sledging might be increased because of cold diuresis and conversely that changes in cold diuresis might be observed in subjects becoming acclimatized to cold. These ideas have prompted studies of water balance and cold diuresis. *Wyatt* [1963], whose work was briefly described earlier [*Edholm and Lewis,* 1964], made careful studies of water balance during sledging journeys. The daily average fluid intake for 157 men was 2560 ± 480 ml, and the urine output averaged 1670 ± 630 ml. There was little consistent difference between active days and rest days, but the patterns of drinking and urine flow were dissimilar. On rest days there was the usual diurnal pattern of urine flow, i.e., low levels at night that rose during the morning to a peak about mid-day; fluid intake, although it was mainly in the morning and evening, was more evenly distributed throughout the day on rest days than during sledging, when nearly all fluid intake accompanied the morning and evening meals. Urine flow during sledging showed two peaks, in the morning and in the evening, low flows occurring during sledging. There was no evidence of dehydration except on occasional days, and cold diuresis could be detected only on about 5% of all days. Wyatt then carried out experiments in which he measured the degree of cold diuresis during a standardized cold exposure in order to answer the question, Is there a diminution of the cold diuretic response during the antarctic winter? The results were inconclusive, partly because of the considerable variation in the response in any one subject.

Recently, *Hillier* [1970] reexamined the phenomenon of cold diuresis, first defining the conditions in which it could be obtained, in experiments carried out in England. The field studies were made at Signy Island, and his subjects included several who were engaged in diving throughout the year and who might have been expected to develop a degree of acclimatization to cold. There was no evidence that the degree of cold diuresis varied during the year or was less in the skin divers than in other subjects. As Wyatt observed, the considerable intra-individual variation may have obscured any tendency for acclimatization effects.

SEBACEOUS GLANDS

The effect of the antarctic environment on the sebaceous glands has been studied by *Corner* [1966], who worked at the Argentine Island base, which is only 65°S, and hence outside activity continues throughout the year. Fourteen men, aged 21–40, were studied, and measurements of sebum secretion from the forehead and the back were made at intervals of 2 weeks from April to November. There was a wide range of variation between subjects; on the basis of the method used, the highest secretor produced 35 times more on the forehead than the lowest secretor. The difference between subjects was smaller when the quantity of sebum secreted on the back was measured, but the difference between the highest secretor and the lowest secretor was sixfold. There was marked consistency in the secretion rate for each individual, and the rank order did not significantly change.

There was a significant fall in the secretion rate with time in the Antarctic for both the forehead and the back, but in the latter case this was due to a considerable reduction of sebum secretion, noted in the last series of measurements, when most of the subjects sunbathed and worked frequently stripped to the waist. Observations were made also of the secretion of sebum on the forehead before and after 10 days of sledging. In eight out of nine subjects there was a decrease, possibly owing to cold or exposure to sunlight or a combination of the two. However, there was no evidence from the main series of observations indicating a seasonal effect. Since the hormone dehydroepiandrosterone is a potent stimulator of the sebaceous glands, it might be expected that an increase of adrenocortical activity would lead to an increased secretion of sebum, but there was no evidence of such an increase and in particular none during the coldest period of the year.

In six subjects, measurements were made in England following return from the Antarctic. There were differences in the individual measurements between the last observations made in the Antarctic and those made in England, but these differences were not consistent, and in the six subjects as a whole there was no significant difference for either the back or the forehead between the antarctic measurements and the English measurements.

Corner's work illustrates once more the dual role of the Antarctic as a laboratory. The opportunity to examine the consistency of individual sebaceous gland activity has been taken, and the results, illustrating not only the very wide individual variation but also the considerable constancy of each individual, are of value in assessing the role of the sebaceous gland. On the other hand, the possible effects of the environment, including the effects of a demanding situation, show that low environmental temperatures have relatively little effect but that solar radiation, as is shown by work elsewhere, can have marked effects.

WOUND HEALING

The effects of a very cold climate on wound healing were investigated by *Catty* [1965], who compared the rate of healing in two groups, one in the Antarctic and the other in England. The final analysis was based on 11 men studied in the Antarctic and nine studied in England. Ages ranged from 21 to 32 years in the antarctic group and from 18 to 26 in the English group.

A standardized wound was made in the forearm at a precisely located site marked with a 1-cm square tattoo pattern and tattoo points 1 cm away along the longitudinal and transverse axes of the forearm. The square of skin was removed aseptically, a full-thickness wound resulting, i.e., with all skin cut from the underlying fat tissue. Tracings of the tattoo and wound area were taken on cellophane, and the wound was covered with a nonadherent dressing. With the arm in a standardized position the wound was examined on the fourth day and then at intervals of 3 days until it was healed. The outline of the unhealed portion of the wound was traced at each examination, as was the position of the dots. Between 5 and 6 months after the wounding the area of the scar was also traced. Although a mild superficial infection of the wound occurred in several subjects, this did not appear to affect the rate of healing. The subjects continued their normal activities and did not seem to be affected in their behavior.

Catty wanted to answer a number of questions: Is wound healing affected by a cold environment? What is the normal pattern of whole thickness skin healing, i.e., time scale, degree of contraction, and direction of contraction? The first question is essentially an antarctic one; the second is concerned with the problems of wound healing as such, irrespective of environment.

On the first question Catty came to the conclusion that there was no essential difference between the two groups, and the implications of this conclusion are discussed briefly below. As far as the series of questions relating to wound healing are concerned, some clear answers were obtained. The tattoo, or wound, area expanded immediately after the skin excision, a greater increase occurring in the transverse axis than in the longitudinal axis. As healing progressed, the tattoo area gradually diminished, but the oblong shape persisted; when healing was complete, the tattoo area was similar in size to the preexcision area. However, 6 months later the scar had increased in size, and the tattoo area was longer than it was at the time of healing. Furthermore, the oblong shape persisted, and the transverse diameter remained longer than the longitudinal diameter. There were some differences between the United Kingdom group and the antarctic group, but these were not statistically significant and may have been influenced by the slight age difference. It is of more interest that the two groups were so similar and that the amount of contraction of the wound after the initial increase accounted for 21–23% of the closure of the wound in both groups. The time scale for healing, of the order of 19 days, was also similar in the two groups.

Catty has quoted laboratory work to the effect that moderately cool environmental conditions (20°C) slow the rate of wound healing in rabbits [*Caldwell*, 1962], but there does not appear to have been any similar study in man. It might be expected that a low skin temperature would diminish the rate of healing, possibly owing to a diminished skin blood flow. The finding that the rate of healing was similar in the Antarctic and in England does not necessarily disprove this view, since it would first have to be established that forearm skin temperature under the conditions of the experiment was significantly lower in Antarctica than in England. Catty did not monitor forearm skin temperature, but it is probable that the difference between forearm ski
temperature in Antarctica and that in England wo
not be large. In England, forearm skin temperature
the subject is comfortable is approximately 33
croft and Edholm, 1943], and skin tempera
trunk is usually close to 35°C. Measureme
clothing temperature, made by using a
mometer woven into the form of a ve
at base the temperature is clos
1960], but during sledging thi
but can seldom fall below

Catty's subjects were stationed at Hope Bay, and, although their activities were not restricted, they did not go on sledging journeys, since they would have prevented making observations. It seems unlikely that average forearm skin temperature would have been more than 1°–2° lower in Antarctica than in England, and such a difference could well be insufficient to have a measurable effect on wound healing. It must be concluded that the answer to Catty's first question is still unsettled.

BACTERIOLOGY AND VIROLOGY

In the course of his studies on wound healing, *Catty* [1965] observed that the incidence of mild infection is higher in England than in the Antarctic and attributed this to the low level of pathogens, as described by *Sladen* [1953]. This statement illustrates a lack of information about the effects of living in an environment that would appear to be virtually free of microorganisms, particularly those producing disease in man.

There have been many descriptions published in accounts of polar explorations about the incidence of upper respiratory tract infections, e.g., the common cold, sore throats, and influenza. The account commonly believed to represent the usual story may be rendered thus. On arrival in Antarctica the members of the base meet the newcomers, and there is an outbreak of coughs, colds, and sore throats. When the relieving ship sails away and leaves the base party in isolation, the upper respiratory tract infections persist but seldom persist for more than 2 weeks. And then until the end of the antarctic year and the arrival of the relief ship there are no more coughs or colds and no more sore throats. It is said that the infection has burned out. On the arrival of the relief ship and contact with the crew and passengers the base members succumb within a few days. Their ...sistance is said to be low, and virtually all catch colds. ...ribed as a legend or even ... evidence to ...scriptions re... ...ficers seldom ... extent.

...ade clear that ...servation than ...rom colds etc. ...fter about 1–3 ...he usual events ... considerably; ...here is quite an ...bers within 2–3 ..., the majority or even all of the wintering party remain quite free from any evidence of infection. On the voyage back to England, some, but certainly not all, of the men from the base catch colds. In a number of cases the first cold is not experienced until 3–4 months after departure from Antarctica, and the statement that the infection is then exceptionally severe is frequently made. Back at base there is usually a mixture of new arrivals and those who remain for a second year, but, as was stated above, upper respiratory tract infections disappear after the second to third week of isolation. One curious variant has been described. The relief ship arrived at Halley Bay, disembarked a number of men, and took on board some 10 men from base who were returning to England. Nearly all 10 men developed colds. The ship had on board also the main party composing the Commonwealth Trans-Antarctic Expedition, who were to be landed at Shackleton base further south in the Weddell Sea. On arrival there (a voyage of some 24 hours), contact was made with the advance party, who had spent a year in the Antarctic, and three men were taken on board while the main party disembarked. During the ship's stay at Shackleton, lasting for about a week, there was contact between the advance party, the main party, and the men from Halley Bay. But none of the Shackleton base advance party caught a cold. This incident illustrates the complexity of the microepidemiological study of the common cold in Antarctica. To obtain a more accurate description, records are being kept by medical officers and others, using a card designed by D. A. J. Tyrrell, Head of the Division of Communicable Diseases of the MRC (Figure 7).

Cameron and Moore [1968] have carried out a study of the epidemiology of respiratory infection at Mawson, an Australian antarctic base. They attempted to isolate viruses from throat and nose swabs and feces collected in Antarctica but were unsuccessful, and they did not find any significant changes in antibody titer for any of the antigens tested. However, the clinical data were of considerable interest in comparison with the anecdotal account given above. The 27 men who formed the group to man Mawson base arrived there in early January after a 10-day voyage from Melbourne. Thirteen of the 27 developed colds during the voyage, which in some cases became worse on arrival at Mawson, but within 2 weeks everyone was symptom free and remained so until the following January. At that time one man spent 3 days with a Russian field party and experienced a mild cold; 18 days later the same man, following a degree of chilling, developed more definite upper respiratory tract symptoms, and so did two other men who had the same experience of body cooling. No one else at base showed

Date Begun.......................... Date Finished.......................... Age........... Name..

Date																												
Day	M	Tu	W	Th	F	Sa	Su	M	Tu	W	Th	F	Sa	Su	M	Tu	W	Th	F	Sa	Su	M	Tu	W	Th	F	Sa	Su
Nose Blocked																												
Nose Running																												
Throat Sore																												
Cough																												
Tight Chest																												
Feeling Tired																												
Headache																												

— = Absent   + = Present, Mild   + + = Present, Severe

Fig. 7. Common cold record card.

any symptoms at all. The relief ship arrived shortly thereafter, and six men developed mild colds during the changeover and on the voyage to Australia. A total of 17 out of 27 developed moderately severe colds, usually within 2 weeks of arrival in Australia.

Recently, some preliminary work was carried out at Stonington Island to assess the feasibility of bacteriological and virological studies in the Antarctic. *Williams* [1969] examined 13 men at approximately monthly intervals for 1 year, swabbing noses, axillae, perinea, umbilici, and palms and examining cultures for the presence or absence of *Staphylococcus aureus.* This organism is present in a high proportion of the population in the nose, for example, and can be responsible for a variety of diseases. Positive cultures were preserved for phage typing in London. Only three subjects were carriers, and they carried distinct strains; there was no evidence of crossover from one subject to another. This lack of cross infection confirmed the extensive investigations carried out by *Sladen and Goldsmith* [1960] and *Sladen* [1965], who examined personnel at a number of U.S. bases during a voyage round Antarctica. The low incidence of *S. aureus* carriers at Stonington Island may be due to the effect of the environment. However, as Williams has pointed out, his technique may have been unsatisfactory, and his results need to be confirmed.

Williams also attempted to inoculate base members with a strain of Coxsackie virus, but this attempt was unsuccessful, probably owing to failure to preserve the virus during the journey south. More recently, a successful experiment was carried out at Stonington Island [*Holmes et al.*, 1971]. Coxsackie virus A_{21} was administered to four subjects, who all developed symptoms. Influenza A_2 virus was administered to another group, but no symptoms were experienced. Coxsackie virus A_{21} apparently also spread to some inoculated men. Nevertheless, fewer symptoms seemed to have been produced in subjects inoculated in the Antarctic than in those inoculated in the United Kingdom.

Another interesting finding was that antibody levels against other respiratory viruses appeared to persist throughout the year spent in isolation. As *Holmes et al.* [1971, p. 197] put it, 'The suggestion that men returning from the Antarctic may experience severe respiratory infection because their antibody levels have fallen is not supported by our results.' Further work is now being undertaken in the Antarctic to confirm the work of Holmes et al.

APPENDIX: BRITISH ANTARCTIC BASES

Adelaide Island (67°46'S, 68°54'W)

Environs. The base is situated on a group of rock outcrops very close to the coastline. To the east and the west the coast continues as high ice cliffs that are part of the margin of the Fuchs ice piedmont. To the north the ice piedmont rises steadily to an elevation of more than 1000 feet and occupies the greater part of the island area. The nearest mountain is Mount Ditte (4600 feet) about 14 miles to the east-northeast. The mountainous part of the island extends from Mount Ditte to the north-northeast, the highest peaks being Mount Liotard (7300 feet), Mount Gaudry (7600 feet), Mount Barré (7200 feet), Mount Bouvier (6800 feet), and Mount Reeves (5380 feet).

Weather. The following conditions may be expected:

Annual mean temperature	−6.1°C (21°F)
Extreme low temperature	−34°C (−30°F)
Extreme high temperature	10°C (50°F)
Annual mean wind speed	14 knots

Complement. Summer, 11–20; winter, 7–12.

Air facility. Snow runway.

History. The base was established as a meteorological station on February 3, 1961. In the following seasons, topographical and geological surveys of the island were made. The station has become the main base for summer aircraft operations in the Marguerite Bay area and to the south.

Scientific disciplines. Meteorology, human physiology, general biological surveys, glaciology.

Buildings. A total of seven one-story buildings stand on rock foundations. The main living hut is T shaped and provides accommodation for 20 men. One section is of prefabricated, insulated timber panels on steel frames (47 feet × 15 feet) with an extension of prefabricated fiber glass laminate (37 feet × 17 feet). In addition, there is a small living hut of prefabricated, insulated timber panels on a steel frame (32 feet × 15 feet) providing accommodation for six men. Normally, this is occupied only in the summer by the air party.

Both the generator hut (32 feet × 15 feet) and the tractor garage (24 feet × 15 feet) are made of prefabricated, insulated timber panels on steel frames. There is also a store (45 feet × 17 feet) made of timber and a small workshop and hut providing sleeping quarters for the duty meteorologist. Anthracite stoves are used for cooking and heating.

Generating plant. Two Lister 8 kw, 230 volt, 50 Hz single phase.

Transport. Two muskeg tractors, dog teams.

Argentine Islands (65°15′S, 64°16′W)

Environs. The Argentine Islands form a small archipelago. Many of them have small ice caps, the highest point in the group being 177 feet on Galindez Island. The distant mainland coast is mountainous and glaciated, the nearest point to the island being Cape Tuxen, 3 miles distant. During the winter, sea ice forms over the whole area, and sledging between the islands and the mainland is possible.

Weather. The following conditions may be expected:

Annual mean temperature	−4.4°C (24°F)
Extreme low temperature	−33.4°C (−28°F)
Extreme high temperature	7.2°C (45°F)
Annual mean wind speed	8 knots

Complement. 11–15.

Air facility. An emergency sea ice runway may be available.

History. The base was established on January 9, 1947, and has existed continuously since then. In 1954 the original building was abandoned, and a new site was occupied at Marina Point, Galindez Island. At the same time, the original meteorological station was expanded to become a geophysical observatory.

Scientific disciplines. Surface meteorology, upper air soundings (radiosonde and radar wind), geomagnetism, ozone, solar radiation, seismology, ionospheric recording, 'whistler' recording, oceanography (tides), glaciology.

Buildings. A total of five one-story buildings stand on rock foundations. The living hut is of timber with steel frames on concrete piers. The whole is clad with insulated timber panels. It provides accommodation for 15 people and space for scientific laboratories. There is also a generator hut (30 feet × 21 feet 6 inches), a nonmagnetic hut (34 feet × 29 feet), a balloon shed, and a shed used for housing meteorological stores. Anthracite is used for cooking, and anthracite and electricity are used for heating.

Generating plant. Two Ford-Petbow 32 kw, 230 volt, 50 Hz single phase; one Lister 12 kw, 230 volt, 50 Hz single phase.

Transport. Dog teams.

Deception Island (62°59′S, 60°34′W)

Environs. Deception Island is one of the South Shetland Islands. It is almost circular, being the top of a volcanic pile emerging above the sea. The main crater wall has been breached by the sea, and it is possible for large ships to enter, through Neptunes Bellows, the extensive sheltered waters within. The base is situated near the site of a deserted whaling station in Whalers Bay. Hot springs abound, and volcanic heat causes the beaches to steam when the tide is out. The airstrip has been constructed on volcanic ash but is covered by snow in the winter. The highest point on the island is Mount Pond (1890 feet).

Weather. The following conditions may be expected:

Annual mean temperature	−2.8°C (27°F)
Extreme low temperature	−10.6°C (13°F)
Extreme high temperature	8.9°C (48°F)
Annual mean wind speed	16 knots

History. This base was established on February 6, 1944, and permanent occupation ceased on December 4, 1967, owing to a volcanic eruption. In February 1969 a second eruption largely destroyed the station and the airstrip. In the future, only short-term summer visits will be made for geological and biological purposes.

Fossil Bluff (71°20′S, 68°17′W)

Environs. The small hut is built at the foot of a rocky slope between the Eros and Uranus glaciers and above George VI Sound, which separates mountainous Alexander Island from Palmer Land. An ice shelf occupies the sound and provides an easy north-south route for traveling parties. In high summer the surface of the shelf becomes flooded with extensive areas of meltwater flowing onto it from the surrounding rocky mountains. To the west and northwest lie the Planet Heights, an extensive range of mountains rising over 5000 feet.

Weather. In the summer the temperature range is likely to be −23°C (−10°F) to 1.7°C (35°F), but, in the winter, temperatures below −40°C (−40°F) may be experienced. Winds in the northern limit of George VI Sound are generally from the north or south, but local katabatic winds descend the glaciers.

Complement. No permanent complement. The hut will house four men and forms an advance station for summer field parties.

Air facility. Unprepared snow runway.

History. This is the only British base to be put in and maintained by air. It was established on February 19, 1961, and parties winter there from time to time.

Scientific disciplines. Survey, geology.

Building. Lightweight prefabricated timber four-man hut with an outer sheathing of aluminium.

Generating plant. One Petter 1.5 kw, 230 volt, 50 Hz single phase.

Transport. Three muskeg tractors.

Halley Bay (75°31′S, 26°42′W)

Environs. The original base was built on the Brunt ice shelf at an elevation of about 100 feet above sea level and some 1½ miles from the seaward edge of the ice shelf; this is known as the ice front. The ice shelf is about 500 feet thick, and, being afloat, it rises and falls with the tide. The surface is slightly undulating, but the general appearance is of a plain. Inland, it is attached to the coastline and merges with the rising inland ice. No rocks are visible from Halley Bay, the nearest exposure being some 200 miles to the south or about 240 miles to the east, the Tottan Hills.

New base buildings were provided at intervals of a few years as the older ones became submerged beneath the accumulating snow. The first hut, built in 1956, is now some 60 feet beneath the surface. In 1967 the present station was built on a new site.

Weather. The following conditions may be expected:

Annual mean temperature	−20.5°C (−5°F)
Extreme low temperature	−51°C (−59°F)
Extreme high temperature	2.7°C (37°F)
Annual mean wind speed	11 knots

Complement. 28–40.

Air facility. None.

History. The base was originally established by the Royal Society in 1955 as a scientific station for the International Geophysical Year and named after Sir Edmund Halley. In 1959 it was taken over by the BAS as a static geophysical observatory. Since then, negotiable inland routes have been pioneered, and topographical, geological, and geophysical surveys are now being made in mountain ranges up to 350 miles distant.

Scientific disciplines. Meteorology, upper air observations (radiosonde, radar wind), geomagnetism, ozone, solar radiation, ionospheric recording, whistlers, aurora and airglow observations (all-sky camera), radio aurora, glaciology, human physiology, survey, geology.

Buildings. In January 1967 the building of a new base was begun about 2 miles further inland than the old station and therefore about 3 miles from the front of the ice shelf. This consists of seven buildings linked by a covered way. Each building (72 feet X 19 feet X 13 feet) is composed of welded steel frames clad with prefabricated, insulated wood panels, which also form the floor. These are extended as the huts are gradually buried by accumulating snow. The shafts also carry all services, chimneys, ventilation, snow chutes for water tanks, aerial feeders, etc. Heating is provided by electric fan heaters to ensure good circulation.

Two buildings provide sleeping accommodation, washrooms, bathrooms, and drying rooms for a maximum of 40 men. One building comprises a lounge, base office, and radio room, and another comprises the kitchen and dining room. A fifth building forms the tractor garage and workshop, and the sixth is the generator house. The last building is the office block containing all the scientific rooms.

The buildings are spaced 20 feet apart so that the snow accumulation between them forms a natural firebreak. Separate structures mounted on skis provide accommodation for the magnetic instruments, ozone spectrophotometer, and other program needs.

Generating plant. Two Rolls-Royce McFarlane 96 kw, 220 volt, 50 Hz single phase (one set is held in reserve to allow rotation for overhaul in the United Kingdom).

Transport. Four muskeg tractors, three International BTD8 tractors, a number of motor toboggans, dog teams.

Signy Island (60°43′S, 45°36′W)

Environs. Signy Island is one of the South Orkney Islands. It is small, being only about 4 miles long and less than 3 miles wide. There is permanent ice over a large part of the island, which attains a maximum height in Tioga Hill (948 feet). The coastline is rocky and has many indentations. In the summer, extensive areas of moss and some tufty grass are exposed, and there are numerous freshwater pools. It is separated from Coronation Island (3000 feet), the largest of the group, by the Normanna Strait, only 1 mile across at the narrowest point.

Weather. The following conditions may be expected:

Annual mean temperature	1.7°C (35°F)
Extreme low temperature	−34.6°C (−31°F)
Extreme high temperature	13.9°C (57°F)
Annual mean wind speed	14 knots

Complement. Summer, 16–21; winter, 13–15.

Air facility. None.

History. The base was established on March 14, 1947, primarily as a meteorological station. Since then the main base hut has been rebuilt twice. In 1964 a combined laboratory and living hut building was built when it became the center for biological work. In the past, topographical and geological work has been done on Signy Island and also on Coronation Island.

Scientific disciplines. Meteorology, soil chemistry and microbiology, nitrogen fixation and photosynthesis of algae, ecology of bryophytes and lichens, biology and chemistry of freshwater and brackish water, physiological investigations of certain crustaceans, survey of sublittoral marine communities (using aqualung diving), ecology and biology of inshore fishes, ornithology, embryology of penguins.

Buildings. A new two-story living hut and laboratory building was built in 1964. The shell is made of panels of prefabricated fiber glass laminate, supported by a steel frame on concrete piers. Each floor is 80 feet 6 inches × 23 feet 6 inches, the upper floor providing living accommodation for 21 people, the lower floor containing five laboratories. The old living hut (84 feet × 24 feet 6 inches) now houses the generating plant, radio office, and workshops and also provides storage space. There is a boat shed, an emergency store, and two small field huts.

Generating plant. Two Rolls-Dale 54 kw, 230/440 volt, 50 Hz three phase.

Transport. One Skidoo motor toboggan.

South Georgia (54°16′S, 36°30′W)

Environs. The station is at King Edward Point in Cumberland East Bay. The island is long and narrow, extending about 100 miles in a northwest-southeast direction and varying in width from about 3 to 25 miles. It is mountainous, with peaks 6000–9000 feet high separated by deep glacier-filled valleys. The land falls steeply to the sea along the southwest coast, where there are no sheltered harbors. The northeast side is also steep, but here there are numerous bays and inlets, around which there is some low-lying ground.

Mount Duse (1663 feet) lies ½ mile to the northeast of the base. Half a mile to the west, across King Edward Cove, is the old whaling station of Grytviken, nestling beneath steeply rising ground. To the south and again across the cove are cliffs 50–100 feet high. From the top of these the ground rises gently to 1000 feet in 1½ miles.

Weather. The following conditions may be expected:

Annual mean temperature	1.7°C (35°F)
Extreme low temperature	−13.9°C (7°F)
Extreme high temperature	23.9°C (75°F)
Annual mean wind speed	7 knots

Complement. Summer, 12–30; winter, 12–25.

Air facility. None.

History. Originally, a government administrative station was set up in 1908. On January 5, 1922, Sir Ernest Shackleton died while his ship lay at Grytviken. In 1925, Discovery Investigations built a laboratory at King Edward Point, maintained for a number of years. On January 1, 1950, the BAS established a meteorological station at the point, and this continued until the work was taken over by the government administration on January 1, 1952. In 1969 the administration was closed down, and the BAS reestablished the present station.

Scientific disciplines. Meteorology, ionospherics, botany.

Buildings. The base consists of one large two-story building, together with a powerhouse, a meteorological office, and various stores. There are also a number of bungalows and other buildings that once served the government administration, but these are no longer in use.

Generating plant. One Mirless National/Brush 150 kw, 230/440 volt, 50 Hz three phase; two McLaren/Brush 75 kw, 230/440 volt, 50 Hz three phase; one McLaren/Brush 27 kw, 230/440 volt, 50 Hz three phase.

Transport. One Massey Ferguson 35 tractor.

Stonington Island (68°11'S, 67°00'W)

Environs. Stonington Island is about 900 yards long and only 400 yards wide. The maximum height is 79 feet. Strictly speaking, it is not an island, for it is joined to the mainland by the Northeast glacier, which passes over its northern extremity. The surrounding mountain scenery is magnificent, and the base has been called the gateway to the south, for journeys can be made over almost unlimited distances on both the east side and the west side of the Antarctic Peninsula. The plateau of the peninsula towers almost 5000 feet above the island, and telltale plumes of wind-blown snow from the summit herald the violent katabatic winds, which occasionally sweep down at speeds up to 100 knots.

Weather. The following conditions may be expected:

Annual mean temperature	–8.9°C (16°F)
Extreme low temperature	–40°C (–40°F)
Extreme high temperature	6.7°C (44°F)
Annual mean wind speed	9 knots

Complement. 13.

Air facility. Emergency snow runway on the Northeast glacier or on sea ice.

History. The base was established on February 24, 1946, to support topographical and geological surveys in the area. It was closed from February 1950 to March 1958 and again from March 1959 to August 1960. Since then, it has been occupied continuously. A new base was built in 1960.

Scientific disciplines. Survey, geology, field geophysics.

Generating plant. Two Lister 6 kw, 230 volt, 50 Hz single phase.

Transport. Dog teams.

Acknowledgments. I wish to express my thanks to the Director of the British Antarctic Survey, Sir Vivian Fuchs, for permission to reproduce the part of the *BAS Handbook* dealing with the descriptions of the individual bases. He has been most helpful also in allowing me to consult and to include material from unpublished reports.

REFERENCES

Antonis, A., I. Bersohn, and D. L. Easty, Serum lipid changes in young men in Antarctica, *J. Physiol. London, 167,* 26–27P, 1963.

Antonis, A., I. Bersohn, R. Plotkin, D. L. Easty, and H. E. Lewis, The influence of seasonal variation, diet and physical activity on serum lipids in young men in Antarctica, *Amer. J. Clin. Nutr., 16,* 428–435, 1965.

Barcroft, H., and O. G. Edholm, The effect of temperature on blood flow and deep temperature in the human forearm, *J. Physiol. London, 102,* 5–10, 1943.

Brotherhood, J., Studies on energy expenditure in the Antarctic, *Polar Human Biology,* edited by O. G. Edholm and E. K. E. Gunderson, pp. 182–192, Heinemann Medical Books, London, 1973.

Budd, G. M., Physiological research at Australian stations in the Arctic and the Subantarctic, in *Human Adaptability to Antarctic Conditions, Antarctic Res. Ser.,* vol. 22, edited by E. K. E. Gunderson, this volume, AGU, Washington, D. C., 1974.

Caldwell, F. T., Metabolic response to thermal trauma, 2, Nutritional studies with rats at two environmental temperatures, *Ann. Surg., 155,* 119–126, 1962.

Cameron, A. S., and B. W. Moore, The epidemiology of respiratory infection in an isolated antarctic community, *J. Hyg., 66,* 427–437, 1968.

Catty, R. H. C., Healing and contraction of experimental full-thickness wounds in the human, *Brit. J. Surg., 52,* 542–548, 1965.

Corner, R. W. M., Sebaceous gland activity of young men in the Antarctic, *Brit. J. Dermatol., 78,* 444–450, 1966.

Davies, A. G., Seasonal changes in body weight and skinfold thickness, *Brit. Antarctic Surv. Bull. 19,* 75–81, 1969.

Davies, T. W., The monitoring of blood pressure during a normal working day using an adaptation of an aneroid sphygmomanometer, *Clin. Sci., 33,* 183–188, 1967.

Davies, T. W., Factors affecting blood pressure studied on man in the Antarctic, M.D. thesis, Univ. of Cambridge, Cambridge, England, 1969.

Davies, T. W., and H. E. Lewis, Simple method for field studies of human blood pressure, *J. Physiol. London, 188,* 7–8P, 1967.

Durnin, J. V. G. A., and R. Passmore, *Energy, Work and Leisure,* pp. 41–44, Heinemann Educational Books, London, 1968.

Easty, D. L., Man in Antarctica, M.D. thesis, Manchester Univ., Manchester, England, 1963.

Easty, D. L., Food intake in Antarctica, *Brit. J. Nutr., 21,* 7–15, 1967.

Easty, D. L., A. Antonis, and I. Bersohn, Adipose fat composition in young men in Antarctica, *Brit. Antarctic Surv. Bull. 13,* 41–45, 1967.

Edholm, O. G., Man and the environment, in *Antarctic Research,* edited by R. Priestley, R. J. Adie, and G. de Q. Robin, pp. 39–60, Butterworths, London, 1964.

Edholm, O. G., Medical research by the British Antarctic Survey, *Polar Rec., 12,* 575–582, 1965.

Edholm, O. G., and A. L. Bacharach (Eds.), *The Physiology of Human Survival,* pp. 3–4, Academic, New York, 1965.

Edholm, O. G., and H. E. Lewis, Terrestrial animals in cold: Men in polar regions, in *Handbook of Physiology,* sect. 4, *Adaptation to the Environment,* edited by D. B. Dill, E. F. Adolph, and C. G. Wilber, pp. 435–446, American Physiological Society, Washington, D. C., 1964.

Elkington, E. J., Finger blood flow in Antarctica, *J. Physiol. London, 199,* 1–10, 1968.

Fry, A. J., Changes in the glucose tolerance test associated with a 'sucrose-free' diet, *Proc. Nutr. Soc., 30,* 63A, 1971.

Hampton, I. F. G., Cold exposure during antarctic journeys, with special reference to its effect on the peripheral circulation in man, Ph.D. thesis, Univ. of London, London, 1967.

Hampton, I. F. G., Effect of cold exposure in the Antarctic on heat elimination from the hands, *Fed. Proc. Fed. Amer. Soc. Exp. Biol., 28,* 1129–1134, 1969*a*.

Hampton, I. F. G., Local acclimatization of the hands to prolonged cold exposure in the Antarctic, *Brit. Antarctic Surv. Bull. 19,* 9–56, 1969*b*.

Hicks, K. E., Changes in the blood-clotting mechanism, serum lipids, and basal blood pressure in Antarctica, *Clin. Sci., 33,* 527–538, 1967.

Hillier, E. R., The effect of dehydration on cold diuresis in recumbent man, *J. Physiol. London, 206,* 18P, 1970.

Holmes, M. J., T. R. Allen, A. F. Bradburne, and E. J. Stott, Studies of respiratory viruses in personnel at an antarctic base, *J. Hyg., 69,* 187–199, 1971.

Lewis, H. E., and J. P. Masterton, Sleep and wakefulness in the Arctic, *Lancet, 1,* 1262–1266, 1957.

Lewis, H. E., J. P. Masterton, and S. Rosenbaum, Body weight and skinfold thickness of men on a polar expedition, *Clin. Sci., 19,* 551–561, 1960.

Lloyd, R. M., Ketonuria in the Antarctic: A detailed study, *Brit. Antarctic Surv. Bull. 20,* 59–68, 1969.

Massey, P. M. O., Acclimatization to cold in Antarctica, M.D. thesis, Univ. of Cambridge, Cambridge, England, 1956.

Massey, P. M. O., Finger numbness and temperature in Antarctica, *J. Appl. Physiol., 14,* 616–620, 1959.

Montegriffo, V. M., Height and weight of a United Kingdom adult population with a review of anthropometric literature, *Ann. Hum. Genet., 31,* 389–399, 1968.

Muir, A. L., Ketonuria in the Antarctic: A preliminary study, *Brit. Antarctic Surv. Bull. 20,* 53–58, 1969.

Natani, K., and J. T. Shurley, Sociopsychological aspects of a winter vigil at South Pole station, in *Human Adaptability to Antarctic Conditions, Antarctic Res. Ser.,* vol. 22, edited by E. K. E. Gunderson, this volume, AGU, Washington, D. C., 1974.

Norman, J. N., Man in the Antarctic, M.D. thesis, Univ. of Glasgow, Glasgow, Scotland, 1960.

Norman, J. N., Cold exposure and patterns of activity at a polar station, *Brit. Antarctic Surv. Bull. 6,* 1–13, 1965.

Orr, N. W. M., Food requirements and weight changes of men on antarctic expeditions, *Brit. J. Nutr., 19,* 79–91, 1965.

Roberts, A. M., Some effects of a sucrose-free diet on serum lipids, *Proc. Nutr. Soc., 30,* 71A, 1971.

Rogers, T. A., and E. G. Aksnes, Studies on arctic survival, in *Aviation and Space Medicine,* pp. 237–240, Universitetsforlaget, Oslo, Norway, 1969.

Rogers, T. A., J. A. Setliff, and J. C. Klopping, Energy cost, fluid and electrolyte balance in subarctic survival situations, *J. Appl. Physiol., 19,* 1–8, 1964.

Rogers, T. A., J. A. Setliff, A. C. Buck, J. C. Klopping, and M. Matter, Ameliorative value of carbohydrate and electrolytes in arctic survival, *J. Appl. Physiol., 21,* 643–648, 1966.

Simpson, H., Stress: Studies in Antarctica, *New Sci., 6,* 927–929, 1959.

Sladen, W. J. L., Bacteriological work in the Antarctic, M.D. thesis, Univ. of London, London, 1953.

Sladen, W. J. L., Staphylococci in noses and streptococci in throats of isolated and semi-isolated antarctic communities, *J. Hyg., 63,* 105–116, 1965.

Sladen, W. J. L., and R. Goldsmith, Biological and medical research based on USS *Staten Island,* Antarctica, 1958-59, *Polar Rec., 10,* 146–148, 1960.

Williams, R. A., Bacteriological survey in Antarctica, *Brit. Antarctic Surv. Bull. 19,* 97–98, 1969.

Wilson, O., Changes in body weight of men in the Antarctic, *Brit. J. Nutr., 14,* 391–401, 1960.

Wolff, H. S., The integrating motor pneumotachograph: A new instrument for the measurement of energy expenditure by indirect calorimetry, *Quart. J. Exp. Physiol. Cog. Med. Sci., 43,* 270–283, 1958.

Wyatt, H. T., Observations on the physiology of men during sledging expeditions, M.D. thesis, Univ. of London, London, 1963.

PHYSIOLOGICAL RESEARCH AT AUSTRALIAN STATIONS IN THE ANTARCTIC AND THE SUBANTARCTIC

G. M. Budd

School of Public Health and Tropical Medicine, University of Sydney, Sydney, New South Wales, Australia

Abstract. The Australian National Antarctic Research Expeditions (ANARE) continuously occupy four stations: Mawson, Davis, and Casey (formerly Wilkes) in Antarctica and Macquarie Island in the Subantarctic. About 80 men winter each year, and a station at Heard Island is intermittently occupied by summer expeditions. Physiological research is done by the medical officer of each expedition following a period of training in Australia or overseas. Thermal stress has been assessed indoors and outdoors. Indoors, men are rarely cold but are often too warm. Outdoors, skin temperature is low despite the clothing worn, and it decreases in the colder weather. Thermal discomfort is common, its incidence averaging 14% at the stations and 31% during travel by dog sledge, and is more frequent in the colder weather and during exercise. Men working outdoors tend to alternate between feeling too hot and feeling too cold, because it is often impractical for them to make the necessary clothing adjustments to compensate for changes in the weather and their activity. Intermittent cold stress is common during everyday life at 'open' stations, where buildings are dispersed and men move between them without donning outdoor clothing. It is suggested that such alternating or intermittent thermal stress could, by exercising the vascular responses to heat and cold, provide an effective stimulus to acclimatization. Acclimatization to cold has been studied by whole-body cold exposures on four Australian and three foreign expeditions. Significant and substantial changes observed in the thermoregulatory responses to such exposures suggest that tissue insulation increases as a result of an enhanced vascular response to cold. Acclimatized men showed a marked decrease in the pressor response to infused noradrenaline. Urinary excretion of adrenaline and noradrenaline did not change with acclimatization, but the excretion of adrenal steroids increased. Changes observed in clothing and thermal comfort have provided indirect evidence of acclimatization. Plethysmographic studies have shown that finger blood flow decreased in the colder months, apparently because of enhanced vasoconstriction. Acclimatization of the face, which is exposed to severe cold stress during outdoor work, has been investigated. Studies of blood coagulation on four expeditions have shown a consistent increase in whole-blood clotting time; possible explanations are that the increase is an accompaniment of acclimatization to cold or that it reflects decreased emotional stress in men who enjoy the simplicity of expedition life. Dietary fat studies have shown that men ate and absorbed much the same amount of fat in Australia and in Antarctica. Standardized continuing observations of body weight, skinfold thickness, blood pressure, arm circumference, heart rate, and oral temperature have been made for more than 10 years. Seasonal changes of the variables, as well as the interrelations between them, have been examined by analysis of covariance; the results have shown considerable variability. It is suggested that the Antarctic provides excellent opportunities for multivariate longitudinal studies.

THE ANARE STATIONS

The Australian National Antarctic Research Expeditions (ANARE), established in 1947 [*Law and Béchervaise,* 1957; *Swan,* 1961], at present (1974) occupy four stations: Mawson, Davis, and Casey (formerly Wilkes) on the antarctic continent and Macquarie Island in the Subantarctic (Table 1 and Figure 1). A station at Heard Island, closed since 1955, was temporarily reoccupied in early 1963, and an inland station on the Amery ice shelf was occupied throughout 1968.

Mawson, Davis, and Casey are coastal stations close to the antarctic circle; hence air temperature rarely falls below −35°C (−30°F), and the limited extent of winter darkness allows outdoor activities, including fieldwork, to continue throughout the year. Wind speed averages 10.3 m/sec at Mawson, 5.8 m/sec at Wilkes, and 4.9 m/sec at Davis. Inland traverses, usually by tractor but sometimes by dog sledge or motor toboggan, are made from Mawson and Casey (and occasionally Davis) each spring and summer and sometimes in autumn and winter as well. They cover distances of up to 2880 km, last from 2 to 4 months, and have encountered air temperatures as low as −63°C. Shorter journeys, often by dog sledge, are frequently made in winter, at Mawson to study the emperor penguin colonies within 80 km of the

TABLE 1. Characteristics of ANARE Stations

Station	Position	Number of Men	Mean Wind Speed, m/sec	Air Temperature, °C Mean	Extreme Maximum	Extreme Minimum
Mawson	67°36′S, 62°52′E	20–30	10.1	−11.2	8.3	−35.4
Davis	68°35′S, 77°58′E	5–9	5.0	−10.6	9.5	−38.3
Wilkes	66°16′S, 110°32′E	20–30	6.0	−10.1	7.2	−37.8
Macquarie Island	54°30′S, 158°57′E	15–20	8.6	4.5	12.4	− 8.9
Heard Island	53°05′S, 73°30′E	9–14	8.3	1.1	14.4	−10.6

Weather data [*Budd et al.*, 1969] are summarized for all years of observation to 1965.

station and at Casey to maintain a glaciological observatory 80 km inland.

Heard and Macquarie islands have typically oceanic wet cold climates, with sea level temperatures near 0°C and a small diurnal and seasonal temperature variation. Wind speed averages 8.5 m/sec, and there are frequent gales. Overcast skies are common, and rain or snow falls on more than 300 days of the year. Macquarie Island lies north of the antarctic convergence, is not glaciated, and is typically subantarctic in character, whereas Heard Island lies south of the convergence, is mountainous and heavily glaciated, and shares characteristics of both subantarctic islands and antarctic islands. On both islands the only practical way to travel during fieldwork is to walk and backpack the necessary equipment and stores. The cold stress of outdoor work is accentuated by the difficulty of keeping clothing dry.

Physiological research at ANARE stations is carried out by the expedition's medical officer following a period of training at the School of Public Health and Tropical Medicine, Sydney. The work is planned, and completed after the expedition, in the school's Department of Environmental Health under the supervision of R. K. Macpherson. Financial support is provided by the Antarctic Division, Department of Supply, Melbourne, from which the expeditions originate.

THERMAL STRESS IN THE ANTARCTIC

Opinion about physiological adaptation in the Antarctic has always varied between two extremes. At one ex-

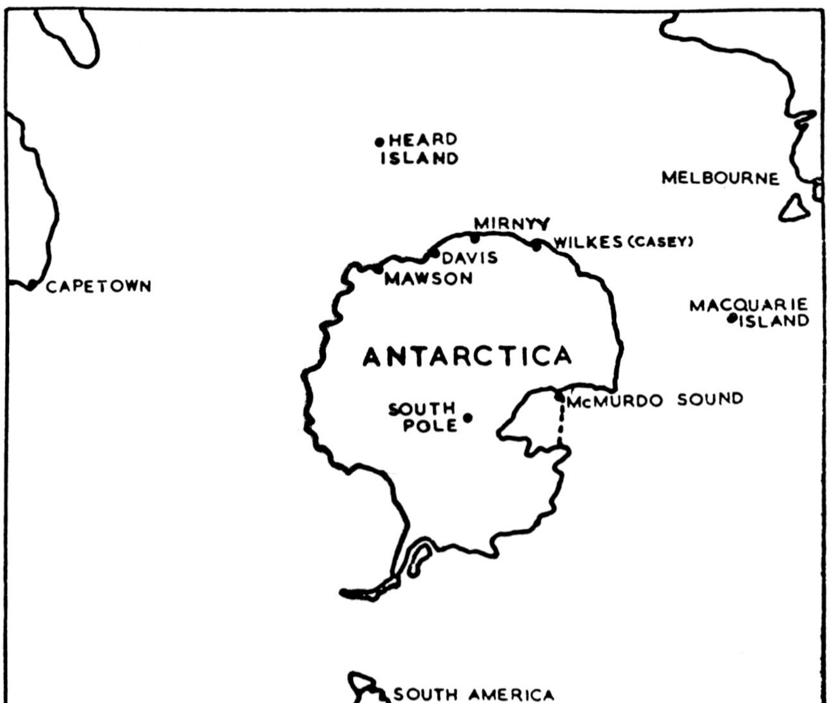

Fig. 1. Location of ANARE stations.

TABLE 2. Indoor Air Temperature and Humidity at ANARE Stations and the Frequency of Thermal Comfort

Station	Year	Observer	Environment		Number of Comfort Votes†	Thermal Comfort, %		
			Air Temperature, °C*	Relative Humidity, %		Too Cold	Comfortable	Too Hot
Mawson	1959	G. M. Budd	20 ± 3	35	179	2 (0)	86	12 (0)
Mawson	1962	D. R. Wigg	15 ± 7	32	480	2 (1)	81	17 (13)
Mawson	1966	J. W. Hudson	16 ± 2	46	151	0 (0)	89	11 (1)
Davis	1962	J. Boda	17 ± 5	44	411	0 (0)	80	20 (6)
Davis	1963	D. J. Lugg	22 ± 4	53	454	2 (2)	86	12 (4)
Wilkes	1962	Z. Soucek	18 ± 2	58	875	0 (0)	98	2 (0)
Wilkes	1963	K. E. Hicks	20 ± 7	41	672	0 (0)	82	18 (6)
Wilkes	1964	L. G. Murray	20 ± 3	46	524	0 (1)	91	9 (7)
Macquarie Island	1963	L. G. Murray	20 ± 5	53	517	0 (0)	83	17 (5)
Mean			18	46		1 (0)	87	12 (5)

Values in parentheses indicate the frequency of shivering (too cold) or of sweating (too hot).
*Values represent the mean plus or minus the standard deviation.
†The total is 4263.

treme there is the naive view that because the climate is cold men must be too, and that practically every physiological change observed is in some way an adaptation to cold. At the other extreme is the view [*Amundsen*, 1912; *Lewis et al.*, 1961] that men are rarely cold because they wear warm clothing and that physiological adaptation is therefore unnecessary. In recent years a number of attempts to resolve this question have been made. They have involved investigations of the thermal stress that men sustain indoors in their heated buildings and outdoors in the cold, and of the way in which they divide their time between these two environments.

Thermal stress indoors. The thermal environment indoors at ANARE stations was assessed over 9 station years (Table 2) by sling psychrometer measurements of air temperature and relative humidity in various buildings, supplemented by estimates of air movement and radiation. Complete records of the clothing worn by the men present were also obtained. The yearly averages ranged from 15°C to 22°C (mean of 18°C) for air temperature and from 32% to 58% (mean of 46%) for relative humidity. At each station, more clothing was worn in the years of lower indoor air temperature.

The yearly averages for air temperature roughly define the 'preferred temperature' for each party, for heating and ventilation are adjusted by the inhabitants of each building to suit themselves. A similar investigation in Sydney [*Hindmarsh and Macpherson*, 1962] showed the preferred temperature of office workers to be 25°C. The preference for lower temperatures in the Antarctic is probably attributable to the heavier clothing commonly worn and the frequent presence of some degree of radiant heat from coal- or oil-fired heaters.

The skin temperature of the abdomen, forehead, and finger of men engaged in sedentary activities indoors was measured on 179 occasions at Mawson in 1959 by a thermistor probe and a null-reading Wheatstone bridge (G. M. Budd, unpublished observations, 1959). Abdomen temperature ranged from 31° to 36°C, but in four fifths of the observations it was 34° or 35°C. This is at or slightly above the upper end of the range (31°–34°C) of skin temperatures considered comfortable by resting men in temperate climates [*Winslow et al.*, 1937], and in fact the frequency of discomfort from heat, which was zero at temperatures below 34°C, increased steadily with increasing temperature above that value. Forehead temperature was distributed in the same way as abdomen temperature but was 1°C lower. Finger temperature, as would be expected, was more variable and ranged from 19° to 35°C, but it was above 30°C in three quarters of the observations, and the most common value was 33°C.

Similar values of skin temperature (mean of 10 points) were observed by *Milan* [1964] at Little America 5; mean skin temperature averaged 32.5°C (range of 32.0°–32.9°C), and finger temperature averaged 28.5°C (range of 21.8°–33.5°C). *Norman* [1965], using a temperature sensitive vest [*Wolff,* 1958], made many measurements of subclothing temperature (the temperature of the space between the clothing and the skin) at Halley Bay and found that it rarely fell below 30°C indoors and was most commonly 33°C.

On each occasion that air temperature was measured indoors at the ANARE stations, the thermal comfort of the men present was also recorded by means of a 7-point 'comfort vote' scale [*Bedford,* 1936]. The scale consists of three major grades: too cold, comfortable, and too hot, which are further subdivided: comfortable into comfortably cool, neutral, and comfortably warm and the extremes into just too hot (or cold) and much too hot (or cold). Subjects rapidly become practiced at describing their thermal sensations in this way. The results (Table 2) show that, although men rarely felt cold indoors, they felt too hot in as many as 20% of the votes and were sweating in as many as 13%. This is a much higher incidence of discomfort from heat than was observed in Sydney [*Hindmarsh and Macpherson,* 1962], where the subjects felt too hot in 4% of the votes and too cold in 2%. It might indicate a habituation to cooler temperatures, but more probably it simply means that the men in the Antarctic did not bother to remove excess clothing when they felt too warm, perhaps because the difficulties of maintaining thermal comfort outdoors (see below) had accustomed them to tolerating some degree of discomfort.

From these observations it appears that when men in the Antarctic are indoors they live in air temperatures similar to those in temperate climates, their skin temperatures are high, and they rarely feel too cold but quite often feel too warm. In other words, cold stress is negligible, but some degree of heat stress is not uncommon.

Thermal stress outdoors. Hourly observations of medial thigh skin temperature, which closely approximates the mean skin temperature of the whole body [*Teichner,* 1958], were made on three men during a 12-day journey of 320 km by dog sledge on the sea ice near Mawson [*Budd,* 1966]. The weather was mild, the mean air temperature being -12°C (range of -17° to -6°C), the wind speed being 6 m/sec (range of 0–36 m/sec), and 5 out of the 12 days being sunny. The exercise, as is usual in sledging, was strenuous, and the average daily energy expenditure on traveling days was estimated from time and motion records to be 5860 kcal (range of 5070–6660 kcal). Skin temperature was 34°–36°C when the men were in their sleeping bags, and at other times when they were inside the tent it ranged from 27° to 36°C, the mode being 33°C. Outdoors, however, it ranged from 18° to 33°C, the mode being 27°C. That is, skin temperature fell by as much as 10°C when the men went outside, in spite of the extra insulating and windproof clothing that they put on at such times. Moreover, skin temperature was lower in the colder weather (Figure 2), even though more clothing was worn. Neither sweating nor variations in activity had any obvious effect.

Very similar results were obtained by *Milan* [1964], who found that heavily dressed men working outdoors at Little America 5 sustained a progressive heat debt and at the end of the observation period (44–165 min) had mean skin temperatures between 26.0° and 28.2°C. Measurements of subclothing temperature by *Norman* [1965] at Halley Bay and by I. F. G. Hampton [*Edholm and Lewis,* 1964] at Hope Bay showed the same temperature distribution, outdoors and in the tent, as those of the men sledging near Mawson and also showed that subclothing temperature fell by 4°–6°C in 10–15 min when the men went outdoors.

Although low skin temperatures suggest that clothing is inadequate and the wearer is cold-stressed, this is not

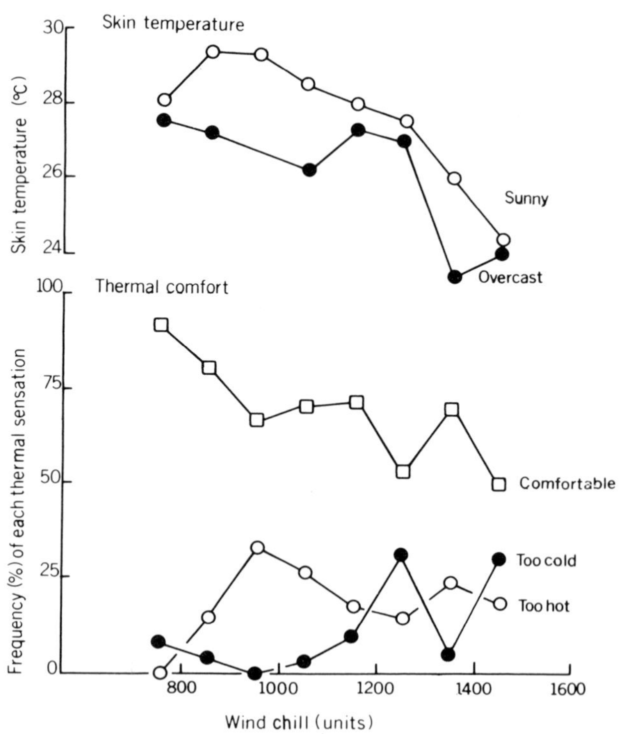

Fig. 2. Skin temperature and thermal comfort of men sledging in Antarctica [*Budd,* 1966]. Wind chill [*Siple and Passel,* 1945] is an index that combines the cooling effects of wind and low air temperature.

always so, for a man with a low skin temperature may feel too hot or too cold, depending on his activity [*Nielsen,* 1938; *Burton and Edholm,* 1955; *Budd,* 1966]. What then is a suitable index of cold stress? Ideally, it would be some measure of thermoregulatory activity, for the object in trying to evaluate thermal stress is to assess the resultant physiological strain, that is, the extent to which the stress evokes the thermoregulatory responses of shivering, sweating, and vasomotor change. Measurement of these responses outdoors in the Antarctic would present many difficulties, but fortunately a reasonable assessment of them can be made from the subject's own report of his thermal sensations. Shivering men feel uncomfortably cold, sweating men usually feel uncomfortably hot, and between these extremes thermal sensation and vasomotor activity both vary with the thermal state of the body [*Burton and Taylor,* 1940]. Whether or not one subscribes to the theory [*Kenshalo and Nafe,* 1963] that thermal sensation originates in the vasomotor activity of the cutaneous blood vessels, the parallel changes commonly observed in these two variables support the view [*Macpherson,* 1962] that thermal comfort or discomfort, properly measured, is a valid index of physiological strain.

In the study of men sledging near Mawson [*Budd,* 1966], thermal comfort of the trunk was reported by means of the comfort vote scale at the time that skin temperature was measured. The results amply demonstrate the difficulty of maintaining thermal comfort during strenuous activity in a cold climate, for the men were uncomfortable in one third of the observations. They also show the important role of heat stress in such situations, for on the average the men were too hot twice as often as they were too cold and were sweating in 23% of the observations. Shivering was recorded only once. Individual differences were striking. One subject was too hot in 41% of the observations and too cold in 8%. Another, who made a deliberate effort to avoid sweating, dressed lightly and adjusted his clothing frequently, the result being that he was uncomfortably cold in 20% of the observations yet was still too hot in 11% and was sweating in 15%.

The limited success of clothing adjustments was also shown by the effects of changes in weather and activity. In the colder weather, thermal discomfort was more common, occurring in as many as 50% of the observations (Figure 2), owing to more frequent discomfort from cold along with little reduction in the frequency of discomfort from heat. Discomfort was also more common during exercise, increasing from 21% of the observations made during light work to 35% of those made during heavy work (Table 3); men were rarely too hot during light work and rarely too cold during heavy work.

Goldsmith [1960] obtained similar results for men working outdoors at Shackleton during the Commonwealth Trans-Antarctic Expedition. Figure 4 of his paper shows that over the year the men felt too hot in 11% of the observations and much too cold (as defined in the ANARE work) in 9%. As the weather became colder, the frequency of discomfort from cold increased from 6% to 13% of the observations while that from heat decreased from 14% to 5%. Thus thermal discomfort of one kind or another remained constant at approximately 20%. These are conservative estimates of discomfort from cold, since I have not included the grade corresponding to just too cold in the ANARE work, owing to uncertainty concerning its correct equivalent on the somewhat different comfort scale used by Goldsmith. The category that probably included it, described by Goldsmith as 'cold extremities,' contained 14% of all votes, and its frequency increased from 6% to 18% as the weather became colder.

The same pattern of thermal discomfort has been observed over several years at a number of ANARE stations (Table 4). Men completed a standard comfort vote questionnaire while they were engaged in routine outdoor work within the station area, usually involving light to moderately strenuous activity. The results of more than 2000 questionnaires, completed by 101 men, show that discomfort of the trunk was due to heat and cold in equal proportion, each occurring in 7% of the observations [*Budd et al.,* 1969]. All the men who were too hot were sweating, but less than half of those who were too

TABLE 3. Effect of Activity on Thermal Comfort during a Sledging Journey

Activity	Number of Observations	Thermal Comfort, %		
		Too Cold	Comfortable	Too Hot
Light work	81	20	79	1
Moderate work	110	9	69	22
Heavy work	85	4	65	31

Data are from *Budd* [1966, Table 3].

TABLE 4. Thermal Comfort of the Trunk in Men Working Outdoors in Antarctica and at Macquarie Island

Station	Year	Observer	Number of Subjects	Number of Observations	Thermal Comfort, %		
					Too Cold	Comfortable	Too Hot
Mawson	1962	D. R. Wigg	24	368	8 (3)	84	8 (12)
Davis	1963	D. J. Lugg	9	473	1 (1)	96	3 (4)
Wilkes	1963	K. E. Hicks	22	387	6 (1)	87	7 (9)
Wilkes	1964	L. G. Murray	23	341	14 (4)	74	12 (12)
Total for Antarctica			78	1569	6 (2)	87	7 (9)
Macquarie Island	1963	L. G. Murray	23	593	7 (3)	87	6 (6)

Values in parentheses indicate the frequency of shivering (too cold) or of sweating (too hot).
Data are from *Budd et al.* [1969].

cold were shivering. The members of one party (Davis in 1963) were unusually successful in keeping comfortable, whereas those of another (Wilkes in 1964) were notably unsuccessful. The average incidence and pattern of thermal discomfort were virtually identical in Antarctica and on Macquarie Island, a finding of some interest in view of the contrast between their climates. In the colder weather, men did not work harder but wore more clothing; nevertheless, the frequency of discomfort from cold increased from 2% to 13% (Figure 3). This finding confirms that of *Palmai* [1962*a*], who in an earlier investigation of thermal comfort at Macquarie Island in 1960 found that there was a significant negative correlation between the mean comfort vote of the trunk and wind chill (an index [*Siple and Passel,* 1945] that combines the cooling effects of wind and low air temperature). Thermal comfort of the head, hands, and feet was also recorded. The hands were uncomfortably cold in 17% of the observations (Table 5); the head, in 12%; and the feet, in 10%. The feet were too hot in 5% of the observations, but discomfort from heat was uncommon for the hands and head. At these sites too, discomfort from cold increased in the colder weather: to 36% of the observations for the head, 33% of those for the hands, and 25% of those for the feet (Figure 3).

Men varied markedly in the kind and degree of thermal discomfort that they experienced (Figure 4). Only 17 of them were always comfortable on the trunk. Of the remainder, 21 were sometimes too hot but never too cold, 22 were sometimes too cold but never too hot, and 41 were sometimes one and sometimes the other. Individual subjects were uncomfortably cold in as many as 59% of the observations for the head, 52% of those for the hands, 45% of those for the feet, and 42% of those for the trunk.

These investigations of skin temperature and thermal comfort show that, although clothing greatly reduces the cold stress of the polar environment, its effectiveness is limited and the limitations become more apparent during colder weather and exercise. Moreover, the reduction in cold stress is often obtained at the cost of an increase in heat stress. Note that the clothing worn was the best that was currently available and that the men were well practiced in its use. The higher incidence of discomfort during sledging than during work at the stations probably reflects the more constant cold exposure, the harder work, and the inadequate drying facilities on such journeys.

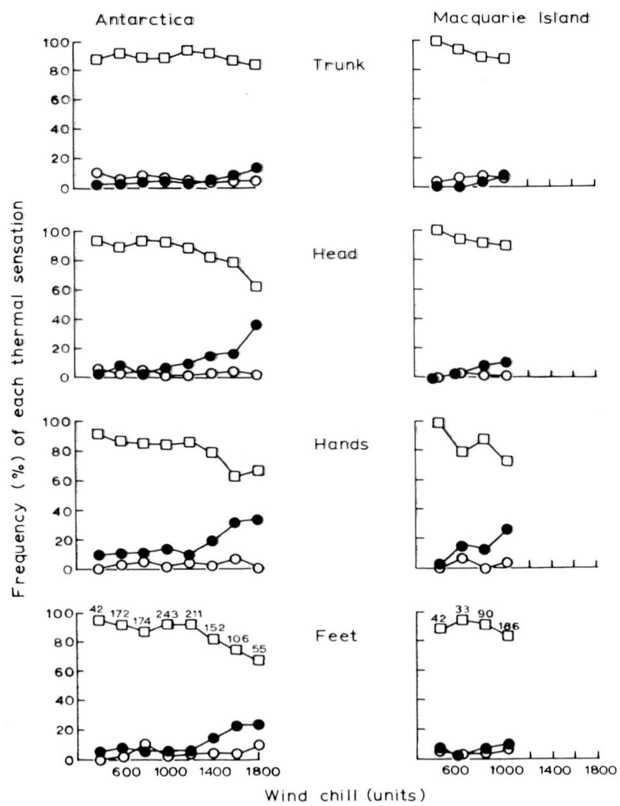

Fig. 3. The distribution of thermal comfort outdoors at each level of wind chill, grouped by intervals of 200 units. Frequencies are expressed as a percentage of the number of observations in each interval, shown next to the symbol for comfortable in the feet diagram. Frequencies based on less than 20 observations were discarded. Squares represent comfortable; filled circles, too cold; and open circles, too hot [*Budd et al.,* 1969].

TABLE 5. Thermal Comfort of the Trunk, Head, Hands, and Feet in Men Working Outdoors in Antarctica and at Macquarie Island

Thermal Category	Thermal Comfort, %			
	Trunk	Head	Hands	Feet
	Antarctica			
Too hot	7 (9)	3	3	5
Comfortable	87	85	80	85
Too cold	6 (2)	12	17	10
	Macquarie Island			
Too hot	6 (6)	1	2	4
Comfortable	87	90	80	89
Too cold	7 (3)	9	18	7

There were 1569 observations for Antarctica and 593 observations for Macquarie Island.
Values in parentheses indicate the frequency of shivering (too cold) or of sweating (too hot).
Data are from *Budd et al.* [1969].

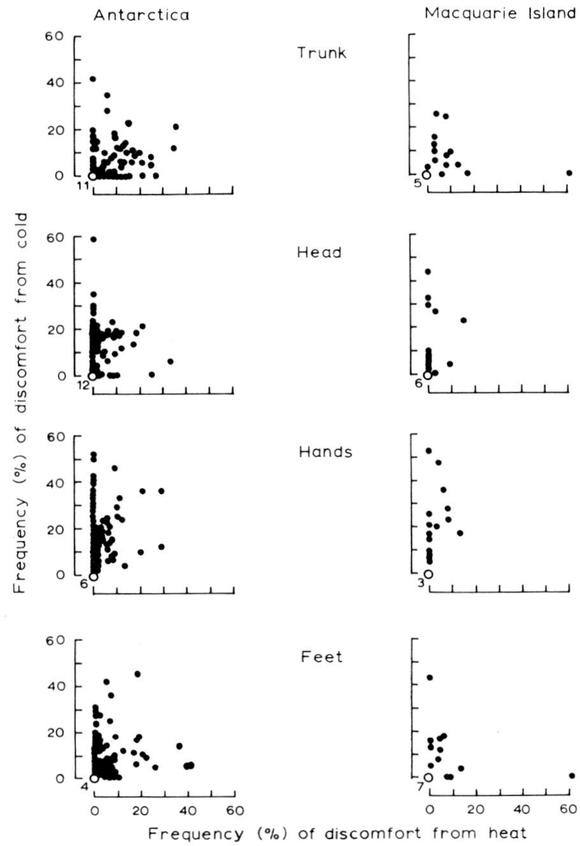

Fig. 4. The frequency of discomfort from heat and from cold in individual subjects. Each point represents a year's observations on one subject, except that the open circle at the origin represents the number of subjects shown by the adjacent numeral. Subjects with less than 10 observations in the year were omitted [*Budd et al.*, 1969].

The results present a paradox, for it might reasonably be expected that men who wear modern polar clothing and intelligently adjust it for changes in weather and activity would be comfortable all the time. The explanation would seem to be that such a view underestimates the difficulties involved. In a warm climate, naked man copes with thermal stress by the physiological mechanisms of vasomotor change, which varies the insulation of the tissues, and of sweating, which is powerful and sensitive enough to handle large and rapidly changing heat loads. In a cold climate, however, the limitations of man's physiological insulation force him to wear heavy clothing, thus placing between his skin and the environment a relatively fixed insulation that can be changed only by conscious activity. This has two unfortunate consequences.

First, the sweating mechanism, normally man's greatest thermoregulatory asset, becomes a liability, for much of the sweat evaporated from the skin during a period of heat stress condenses in the cold outer layers of the clothing. This condensation reduces the efficiency of cooling by sweat to as little as 40% of its normal value [*Belding et al.*, 1947] and also serves to wet the clothing and so reduce its insulation. Thus sweating not only fails to keep man cool when he is active but also tends to defeat his subsequent attempts to keep warm when he is inactive.

Second, the more clothing worn, the smaller becomes the fraction of the total insulation that is under physiological control. The burden of thermoregulation is progressively shifted from man's autonomic mechanisms to his voluntary activity, and the maintenance of thermal comfort becomes a conscious and time-consuming art, in which the likelihood of success depends on the amount of time and patience that a man is able to devote to it. Thermoregulation is thus in competition with the other tasks that have a claim on his time, and it is perhaps surprising that he achieves even as much success as he does. Every change in activity or weather (particularly wind and sunshine) changes the clothing requirement, but often it is simply not convenient for him to stop what he is doing so that he can adjust his clothing, especially if he is engaged in group activities such as sledging or roped travel or if the weather is bad. To change a sweater he must first halt and remove his outer garments, together with such encumbrances as rope, pack, or camera. In bad weather he gets snow in his clothing while he is changing, and even his ability to ventilate his clothing is limited because snow (or rain in the Subantarctic) enters through the ventilation openings. At such times the alternative to getting wet from sweat is getting wet from external water, or else, if he deliberately underdresses to avoid sweating, putting up with feeling too cold for much of the time.

These difficulties lead him to postpone making clothing adjustments and instead to tolerate moderate degrees of thermal discomfort. In most cases he dresses so as to be too warm rather than too cold, partly because he is more used to warmth from his previous experience but also because sweating does not interfere with his immediate activities in the way that shivering or numb and painful extremities do, even though it makes these consequences more likely when he later reduces his level of activity.

The classic accounts of polar travel [*Nansen,* 1895; *Cherry-Garrard,* 1952; *Wilson,* 1966] abound in descriptions of heat and profuse sweating on the march, alternating with severe chilling during halts. The men's discomfort was probably due not to ignorance of the role of sweating in exercise but to the impossibility of their making any effective clothing adjustments when they were wearing a sledge-hauling belt over a windproof parka and hauling for long periods with other men of differing thermal tolerance. The same pattern of overheating during sledging and chilling during rest halts was observed at Mawson [*Budd,* 1966] (although the chilling, coming as it did a few minutes after each hourly observation, was not recorded) and is in fact a commonplace of strenuous exertion in cold weather. Admittedly, the modern polar traveler (and even more the man working outdoors at the station) usually manages to avoid the extremes of discomfort suffered by his predecessors, mainly because he does not work so hard, because he can adjust his clothing more easily, and because he generally has enough fuel to dry his clothing to some extent. Nevertheless, his basic problems of thermoregulation remain the same. Far from being thermally isolated by his clothing, he is quite often in an unstable equilibrium between being too hot and being too cold, and the harder the work or the colder the weather, the more unstable the balance becomes.

Time spent outdoors. The amount of time that men spend outdoors in the Antarctic has been estimated on several occasions. Time and motion studies were made at three stations of the British Antarctic Survey; two of them (Halley Bay and Argentine Islands) were 'static bases,' where little fieldwork was being done, and the other (Stonington Island) was a 'sledging base.' At each station the observer recorded in detail the activities of four men for 24 hours each month of the year. The average time spent outdoors over the year was 2.2 hr/day at the two static bases [*Norman,* 1965; A. Cumming, *Edholm and Lewis,* 1964] and 3.6 hr/day at the sledging base [H. T. Wyatt, *Edholm and Lewis,* 1964]. A marked seasonal variation was observed at Halley Bay, where the subjects spent 1.2 hr/day outdoors in winter and 3.1 hr/day outdoors in spring.

Similar results were obtained at the U.S. station Little America 5 by a different technique [*Milan and Rodahl,* 1961]. Activity cards were compiled by groups of subjects for 5-day periods in autumn, winter, and spring. The time spent outdoors by four scientists in these three seasons was 3.1, 1.4, and 3.6 hr/day, respectively, the overall mean being 2.6 hr/day. A group of four sailors acting as support personnel were outdoors for 4.8 hr/day in the spring.

At the ANARE station Wilkes a more comprehensive survey was made by a less accurate method [*Soucek,* 1963]. On one afternoon each week the observer questioned every man at the station about his activities for the day, and the next morning he completed the interview, thus covering a 24-hour period. The results of 817 such interviews showed that the average estimate of time spent outdoors was 2.4 hr/day. Outdoor time varied with men's occupations rather than with the seasons: a physicist averaged 0.9 hr/day, and an electrician averaged 3.7 hr/day. These values include the time spent in unheated shelters but do not include the time spent outdoors by men who were away on field trips at the time of the interviews.

The same technique was used by Soucek and D. Foster [*Soucek,* 1963] to estimate the time spent outdoors by six members of the same party during a tractor journey of 2880 km to the Soviet inland station Vostok. The results of 12 weekly interviews during the 4-month journey showed that the men spent 3.2 hr/day outside their warm tractor cabins or caravans, as compared with their 2.9 hr/day for the remainder of the year. Air temperature during the journey averaged −31°C (minimum of −63°C), and wind speed averaged 4.5 m/sec.

On sledging journeys, more time is necessarily spent outdoors. Detailed time and motion studies showed that men were outdoors for 9.4 hr/day during a 7-day man-hauling journey from Halley Bay [*Norman,* 1965] and for 9.2 hr/day during a 12-day dog-sledging journey from Mawson [*Budd,* 1966]. Even when they are indoors in their tent, men who are sledging are in a cold environment except when meals are being cooked on the primus stove, and they usually have to get into their sleeping bags to keep warm.

Although methods varied, these studies agree in showing that men working at the stations spent some 2.2–3.6 hr/day outdoors. These times are equivalent to 9–15% of each day, or 14–23% of the waking hours, that is, of the time available for outdoor activities. The sledging studies also agree in showing that men spent an average of 9.3 hr/day outdoors, equivalent to 39% of each day, or 56% of the waking hours.

A crude estimate of the duration of thermal stress can now be made by multiplying the average time spent outdoors by the average frequency of thermal discomfort. At the stations, men were too cold for 7% of the outdoor time of 2.5 hr/day (as a representative figure), which amounts to 10.5 min/day. They were too hot and sweating for the same length of time. During sledging, men were too cold for 11% of the 9.3 hr/day spent outdoors, that is, for 61 min/day. They were too hot for 20% of the time, that is, for 112 min/day, and were sweating for 128 min/day.

Rearranging these estimates, we find that men felt too hot for 11 min/day at the stations and for 2 hr/day during sledging. They felt too cold for 11 min/day at the stations and for 1 hr/day during sledging.

Differences between stations. Modern antarctic stations vary greatly in their latitude, their design, and the activities pursued, and these factors can profoundly influence the thermal stress sustained by their inhabitants [*Budd,* 1964*a*]. The contrast between the thermal stress of sledging and that of station life underlines the importance of the distinction [*Edholm and Lewis,* 1964] between static bases and sledging bases. A distinction that may be equally significant is that between 'compact' stations and 'open' stations [*Law,* 1965].

Compact stations consist of either a single general-purpose building or a number of buildings connected by weatherproof passageways. Open stations consist of a large number of special-purpose buildings dispersed over a wide area. Most present-day stations are compact, especially those built on ice or névé, which inevitably become buried under the snow. They provide easy access to all facilities in any weather but are much more vulnerable to fire than the dispersed buildings of open stations. Examples of open stations in Antarctica are McMurdo, Mirnyy, and Mawson, each of which consists of 50 or more buildings scattered over a radius of more than 400 meters.

Of the stations where the physiological work considered here has been done, Argentine Islands, Hope Bay, and Stonington Island (all of them built on rock) are compact, in keeping with the traditional British preference for a single large 'base hut.' Halley Bay, Sanae, and Little America 5, built on ice shelves, were soon buried many meters beneath the surface. Shackleton at the time of *Goldsmith*'s [1960] observations was still under construction by men who lived in tents and a tractor crate and were therefore in an open station situation. As for the ANARE stations, Wilkes and Casey (the replacement station near Wilkes, occupied in 1969) are compact stations consisting of many buildings linked by passage ways; Davis is semi-open, for the main work huts and store huts are 30–200 meters from the interconnected group of living huts; Macquarie Island is open, many of the work huts being 200–400 meters from the group of living huts, which are themselves 10–30 meters apart; and Mawson, described in detail [*Budd,* 1964*a*] before it grew to its present size, is the most open station of all. Note that so far all estimates of the time men spend outdoors have been made at compact stations.

These differences in station design are physiologically significant. At a compact station, most men can stay indoors for days or weeks at a time if they wish and can postpone outdoor excursions until a period of fine weather [*Norman,* 1965]. At an open station they are exposed to the weather every day of the year and in strong winds must haul themselves along a stout rope linking the buildings.

Several consequences follow. First, men at open stations spend more time outdoors. At Mawson, which well exemplifies conditions at an open station, even men whose work is done indoors must make at least a dozen excursions between huts each day and commonly 20 or more. Since each excursion lasts only a minute or two, the additional outdoor time might not amount to much more than 30 min/day, the total for the day thus increasing from the 2.5 hours of a compact station to 3.0 hours. If the frequency of discomfort during these excursions were similar to that during outdoor work, the additional cold stress, 7% of 30 min, would raise the daily total from 11 min to 13 min–hardly a very striking increase.

However, the frequency of discomfort during these excursions is not similar to that during outdoor work, for the men dress differently. This difference is the second consequence of living in an open station. The thermal comfort survey described above was made on men who, working outdoors for 30 min to several hours, were wearing the outdoor clothing appropriate to an exposure of that duration. By contrast, men moving between the huts at Mawson do not as a rule bother to put on outdoor clothing unless a blizzard is blowing–it is too much trouble when one has to do it 20 or more times a day–and consequently they become superficially chilled from the constant wind, which rapidly penetrates their indoor clothing. They generally feel too cold for a few minutes and may even shiver a little, but they soon rewarm. The total duration of discomfort from cold may thus approach 60 min/day, and the duration of vasoconstriction may be longer still. Measurements made immediately before and after a few of these brief exposures during quite mild weather at Mawson (a wind of 10 m/sec at a temperature of -7°C) showed that medial thigh skin temperature always fell by 2°C or more and once fell to as low as 24°C. Moreover, some degree of vasoconstriction, as shown by persistently low temperatures of the thigh and finger, lasted for 10–30 min, although discomfort was usually short-lived.

The third consequence is that the pattern of thermal stress at an open station is different from that at a compact station. Instead of a few sustained periods of mild cold stress, there are many brief periods of quite severe cold stress, endured precisely because they are so brief. In an exposure of more than a few minutes this cold stress would be intolerable. The vasomotor response to cold is thus repeatedly activated by abrupt, intense, and short-lived stimuli, which vary in their intensity (but not their frequency or duration) with the outdoor weather, rising to a maximum in the winter and then declining.

Stimulus to acclimatization. The work so far considered appears to show that the truth about cold stress in the Antarctic lies somewhere between the two extreme views quoted at the outset. Man is not under continuous or severe cold stress, for he gains a great deal of protection from his housing and clothing. But neither is he thermally isolated from the environment, for a certain amount of cold stress remains.

Is this residual cold stress enough to induce acclimatization? Its average daily duration and severity appear to be much less than those required to acclimatize rats in the laboratory and also much less than those used in many unsuccessful attempts to acclimatize man in cold rooms. They compare more favorably, especially during sledging, with the duration and severity of the exposures of 50 min/day that have proved adequate to acclimatize man to heat [*Lind and Bass,* 1963], but to this comparison the objection could be raised that these are 50 min/day of the appropriate stimulus to heat acclimatization, namely, an elevated sweat rate, which in due course leads to the enhanced sweating response that characterizes acclimatization [*Fox et al.,* 1963]. It might further be argued that the appropriate stimulus to cold acclimatization in laboratory rats is continuous shivering, which is gradually replaced by the nonshivering thermogenesis and constantly elevated heat production that characterize the acclimatized rat. Since man in the Antarctic is rarely cold enough to shiver, how could he acclimatize?

The answer would seem to be that he could not, if he resembled the rat exposed to constant cold in the laboratory. But the resemblance, as is pointed out by *Wyndham et al.* [1964], is rather to a different rat. In his pattern of exposure and his adaptation to it, man seems more like the rat exposed to seasonal cold outdoors [*Héroux et al.,* 1959]. His exposure is variable rather than constant, and he adapts by increasing his tissue insulation rather than his heat production.

The characteristic thermal stress in the Antarctic is intermittent periods of cold stress and heat stress. Thermoregulatory responses are thus called on to vary over a wide range at frequent intervals, alternately conserving heat by vasoconstriction and dissipating it by vasodilatation and sweating. Colder weather increases both the frequency of the episodes of chilling during outdoor work and the rapidity of chilling during the brief exposures of station life. The most likely adaptation to this kind of 'vasomotor training' would seem to be an increase in the sensitivity and the intensity of the vasomotor responses, resulting in a more prompt and effective adjustment to thermal stress of either kind. Evidence from Mawson and Heard Island, described in the next section, indicates that men's responses to cold do change in just this way, and recent observations [*Wilkins,* 1973] suggest that similar changes occur in their responses to heat.

ACCLIMATIZATION TO COLD

Direct evidence of general acclimatization. The most direct way to seek evidence of general acclimatization to cold in the members of polar expeditions is to expose them periodically to cooling of the whole body and see whether their thermal and metabolic responses change. This has been done on at least seven expeditions to the Antarctic.

The first Australian study by this method was made at Mawson in 1959 [*Budd,* 1962, 1964*a*]. The results provide clear evidence of acclimatization and will therefore be described in some detail. Four members of the 1959 Mawson party were exposed to a standard test cold exposure before, during, and after their year in Antarctica. In each test they were exposed to air at a temperature of 10°C for 95 min while they lay naked on a mattress of nylon tennis net. Five series of tests were done, and, in every series, each subject was exposed twice, so that there were eight exposures per series and 40 for the whole study. Series 1 was done in Melbourne in December (summer) 1958 immediately before the subjects sailed for Antarctica; series 2, 3, and 4 were done at Mawson in 1959 in the autumn, spring, and summer, respectively; and series 5 was done a week after the return to Melbourne in March (autumn) 1960, 1 month after the expedition ship had left antarctic waters. During the test cold exposure and for the last half hour of a 1-hour control period under blankets that preceded it, skin temperature at five sites and rectal temperature were measured (with thermistors and a null-reading Wheatstone bridge) every 5 min, and the oxygen consumption was determined (by an open circuit method using Douglas bags and a Scholander gas analyzer) for 10 min out of every 15. Shivering was assessed by frequent visual inspection, and the air temperature, humidity, air movement, and mean radiant temperature were measured during every test by standard methods.

The main finding was that the ability to maintain rectal temperature improved at Mawson (Figure 5). In series 1 the test cold exposure caused an initial fall in the average rectal temperature, followed by a slight rise and then a secondary fall. In series 3 there was hardly any initial fall, and the rise was early, high, and well sustained. In series 5 a slight early rise occurred but was not sustained, and the rectal temperature fell steeply to a lower level than that in series 1. In both pattern and degree, series 2 was intermediate between series 1 and series 3, and series 4 was intermediate between series 3 and series 5.

This general relation between series held for every subject, but there were systematic individual differences in the pattern of response and in the degree to which the subject succeeded in maintaining his rectal temperature

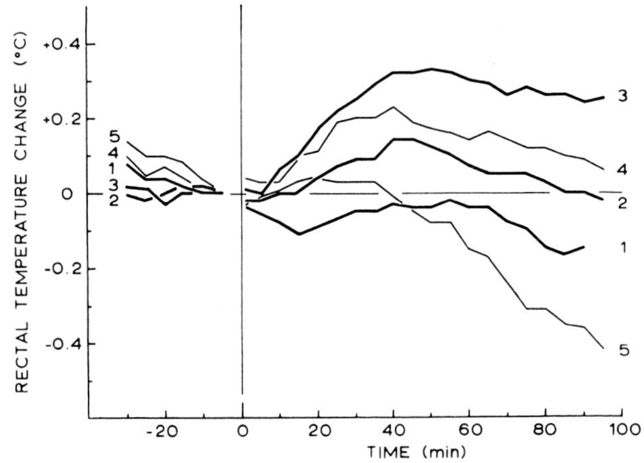

Fig. 5. Rectal temperature change in Melbourne and at Mawson. Mean values are given for the eight exposures in each series. The numbers adjacent to the curves identify the series. Zero time indicates the commencement of the cold phase [*Budd,* 1964*a*].

(Figure 6). The ability to maintain body temperature was apparently related to body size, the heavier and taller subjects (subjects M, B, A, and S in descending order) being the more successful. However, changes in body weight throughout the study could not explain the changes in rectal temperature response, for the two vari-

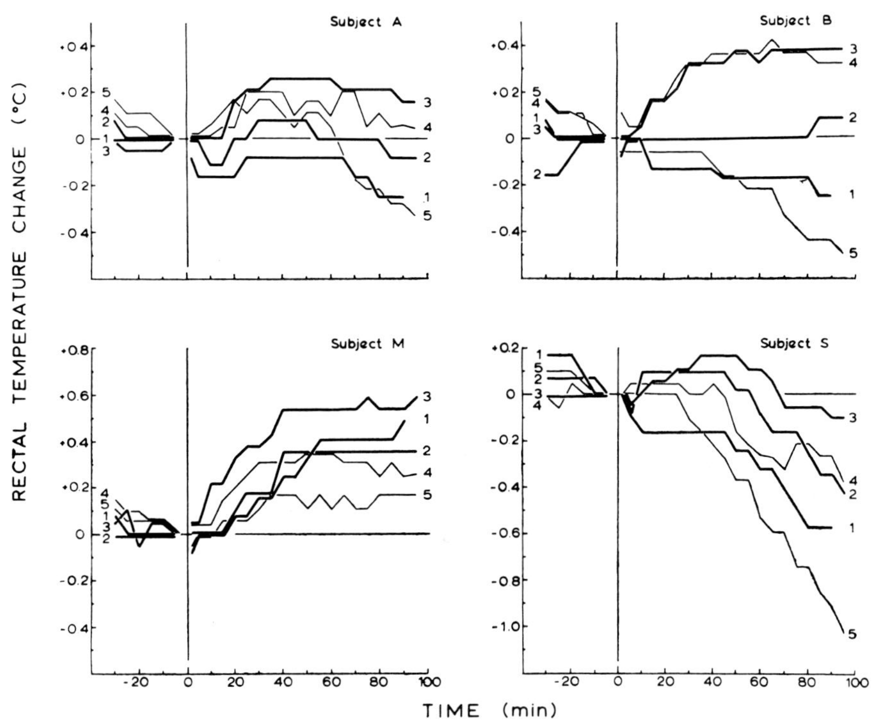

Fig. 6. Rectal temperature change in Melbourne and at Mawson. Individual values (mean of two replicates) are given for each series. The numbers adjacent to the curves identify the series. Zero time indicates the commencement of the cold phase [*Budd,* 1964*a*].

ables were not significantly correlated ($P > 0.20$): subject B, for instance, showed striking changes in his rectal temperature response (Figure 6) but weighed less at Mawson than in Melbourne. Nor were the changes in rectal temperature related to minor variations in the test cold stress, to changes in relative humidity, or to the subjects' assessments of their physical fitness. On the other hand, there appeared to be a clear relation to air temperature (Figure 7), and it was therefore concluded that the changes observed at Mawson were the result of general acclimatization to the cold.

As regards the mechanism of acclimatization, it was certainly not mediated by increased heat production. Oxygen consumption immediately before the cold exposure remained constant in all series, and the increase caused by the exposure was unrelated to the final rectal temperature. No early elevation of oxygen consumption occurred to explain the acclimatized man's early rise of rectal temperature; on the contrary, it was often lower

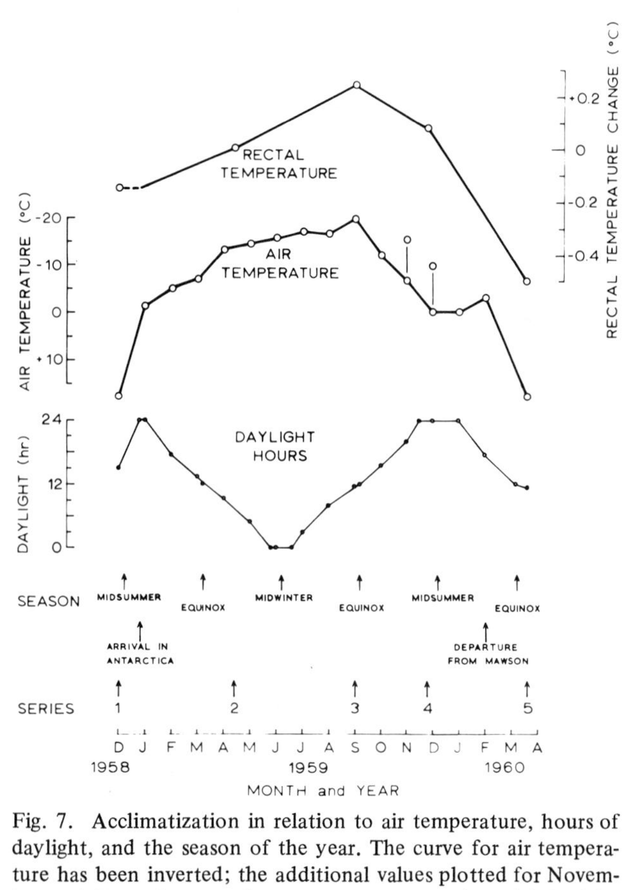

Fig. 7. Acclimatization in relation to air temperature, hours of daylight, and the season of the year. The curve for air temperature has been inverted; the additional values plotted for November and December are the temperatures experienced inland by subjects A, B, and S. The hours of daylight, taken from current astronomical tables, are those between sunrise and sunset. Rectal temperature in series 5 has been adjusted for an anomalous response in one exposure [*Budd,* 1964*a*].

than that of the unacclimatized man for the first half hour of the cold exposure, during which time rectal temperature was nevertheless rising steeply. As occurred in the case of rectal temperature, each man had his own characteristic pattern of metabolic response (Figure 8). There was no evidence of 'nonshivering thermogenesis,' for the onset, progression, and severity of shivering paralleled that of the increase in oxygen consumption.

Improved maintenance of rectal temperature without a corresponding increase in heat production would seem to imply increased tissue insulation, and temperature measurements during rewarming confirmed that the subjects had lost less heat when they were acclimatized than when they were unacclimatized. That the increased insulation resulted from vascular changes rather than from fat deposition was suggested by the absence of correlation with weight changes and by the nature of the rectal temperature response. Paradoxical rises of rectal temperature are attributable to peripheral vasoconstriction [*Aschoff,* 1944; *Bazett et al.,* 1948; *Grayson,* 1950, 1951], and the early, high, and sustained rise of rectal temperature in the acclimatized man was therefore taken to indicate a more prompt, intense, and sustained vasoconstriction than had occurred when he was unacclimatized. Observations of synchronous changes of skin and rectal temperature support this interpretation. It was suggested that the promptness of the vasoconstriction avoided the loss of heat from the deep body tissues that would otherwise have occurred between the onset of cold exposure and the establishment of effective vasoconstriction, and that its greater intensity and duration served to reduce the subsequent rate of heat loss.

These changes were well illustrated by the response of subject S in series 2 (Figure 9), in which there was a prompt rise in rectal temperature, a very late response of shivering and increased heat production, and the lower skin temperature that enhanced vasoconstriction would seem to imply. However, such a clear pattern was rarely seen, especially for skin temperature, although extremity temperatures were generally lower at Mawson than in Melbourne. Paradoxical results for skin temperature in short cold exposures have also been reported by other observers [*Adams and Covino,* 1958; *Heberling and Adams,* 1961; *Milan et al.,* 1961; *Davis,* 1963; *Wyndham et al.,* 1964] and are probably attributable to the fact that thermal equilibrium has not been attained.

Incidental findings of this study were that there were no consistent changes between series in cold diuresis, heart rate, and thermal discomfort.

The investigation was repeated and extended at Mawson in 1964 [*Budd and Warhaft,* 1966*a*]. Four subjects underwent test cold exposures in exactly the same

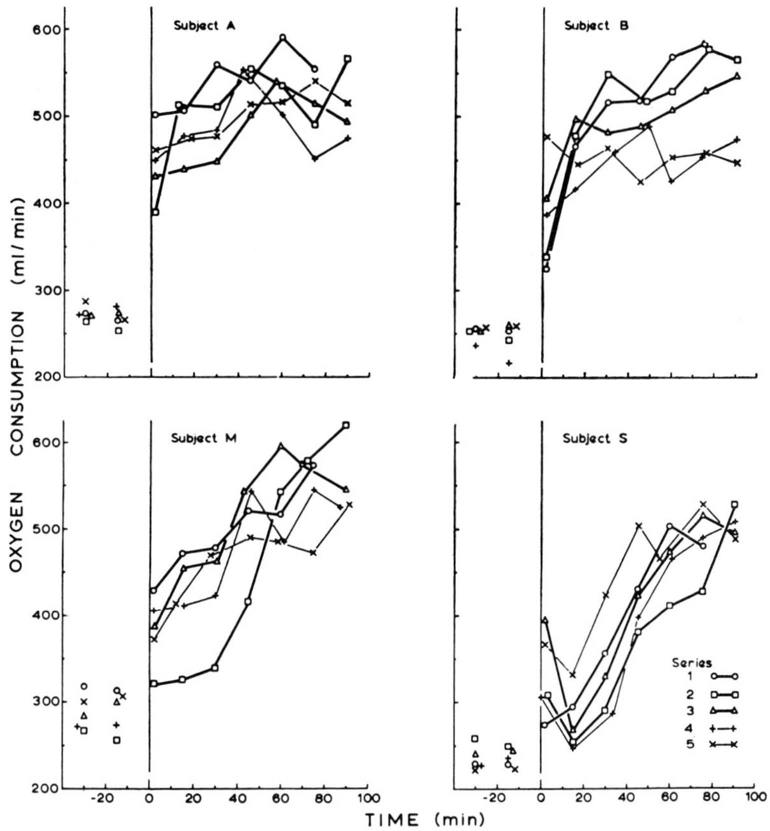

Fig. 8. Oxygen consumption. Individual values (mean of two replicates) are given for each series. Zero time indicates the commencement of the cold phase. Each value is plotted against the time of commencement of the 10-min period during which it was determined [*Budd,* 1964*a*].

manner as was used in 1959, but the exposure lasted for 120 min instead of 95. Two complete series, each consisting of eight experiments, were done, one immediately before the subjects sailed from Melbourne and another at Mawson in late winter 24 weeks after arrival. A third series was begun in early spring but was only half completed.

The results confirmed the 1959 findings. Rectal temperature changed in just the same way (Figure 10), and the rate at which the change developed per week in Antarctica was virtually the same. Temperature measurements during rewarming support the view that acclimatization is mediated by enhanced vasoconstriction, and so do certain abrupt fluctuations in rectal temperature that occurred during the exposures (Figure 11 and also Figure 6). Shivering and cold diuresis did not change, nor did skin temperature except for a small increase in toe temperature. Bradycardia caused by the cold exposure was greater at Mawson than in Melbourne, but the rise in blood pressure was the same as it had been in Melbourne. As occurred in 1959, the changes in rectal temperature response were not correlated with those in body weight. The results of the incomplete third series closely resembled those of the second series.

An anomalous response to one of the test exposures in 1959, by a subject suffering from lack of sleep, suggested that acclimatization might be mediated by a change in the production or utilization of the catecholamines adrenaline and noradrenaline [*Budd,* 1964*a*]. Urine samples were therefore collected before and during the cold exposures in each series of the 1964 investigation and later analyzed in Australia [*Budd and Warhaft,* 1970]. During the cold exposures, significant increases occurred in the excretion of 17-hydroxycorticosteroids, 17-ketosteroids, adrenaline, noradrenaline, sodium, and potassium. The increase was the same at Mawson as it had been in Melbourne, and catecholamine excretion in both the preexposure period and the exposure period did not differ significantly between the two places. Steroid excretion, however, was significantly greater at Mawson than in Melbourne in both the preexposure and the exposure period, and to a similar extent

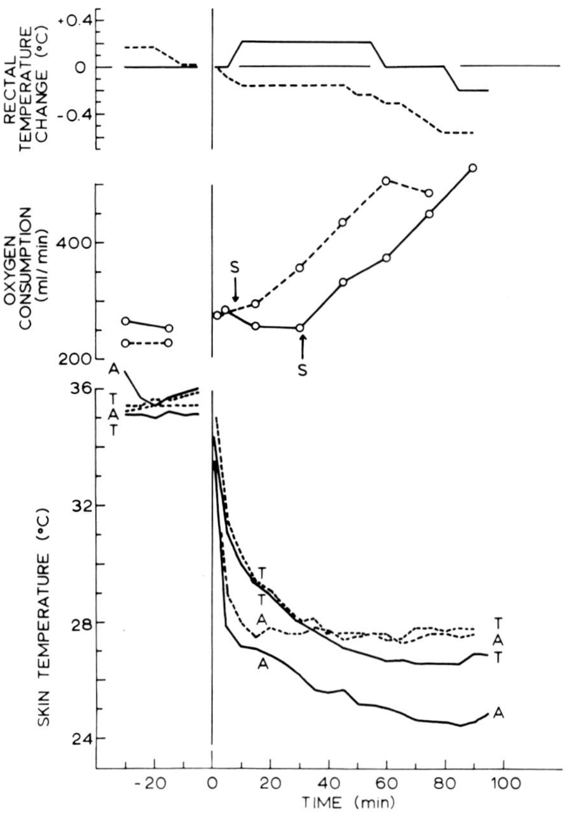

Fig. 9. Body temperature, oxygen consumption, and onset of shivering of subject S in the second exposure of series 2 compared with his mean values in series 1. Zero time indicates the commencement of the cold phase. The solid lines represent series 2*b*; broken lines, series 1; A, abdomen; T, thigh; S, the onset of shivering [*Budd,* 1964*a*].

for the 17-hydroxycorticosteroids and 17-ketosteroids. The increased steroid excretion resembled that found in men working at McMurdo Sound in early summer [*Pace,* 1960] but contrasted with the reduced excretion observed by *Staquet* [1961] at Roi Baudouin station in 1959. The unchanged excretion rates of adrenaline and noradrenaline suggest that, if catecholamine production had changed during acclimatization at Mawson, it must somehow have been balanced by changes in utilization.

This possibility was emphasized by the results of an additional study made in 1964 [*Budd and Warhaft,* 1966*b*]. Within a few weeks of the test cold exposures in Melbourne and at Mawson, the same four subjects were infused with noradrenaline at four dosage levels (5–22 μg/min), and their cardiovascular and metabolic responses were measured. The most remarkable finding was that the pressor effect of noradrenaline, as well as the accompanying symptoms, decreased markedly at Mawson in three of the four subjects (Figure 12). This reduced sensitivity of the blood vessels to noradrenaline evidently did not include the skin vessels, because the effect of the infusion on skin temperature either increased or remained unchanged.

One aim of the study was to see whether a calorigenic response to noradrenaline developed with acclimatization at Mawson, as it does in acclimatized rats [*Hsieh and Carlson,* 1957; *Depocas,* 1960], and, to a limited extent (18-ml/min/m^2 increase in oxygen consumption), in men exposed in cold rooms [*Joy et al.,* 1963]. The results are inconclusive because the masks used to collect expired air leaked somewhat around the subjects' beards, but they appear to show that oxygen consumption, which in Melbourne was not affected by noradrenaline, increased in proportion to the dose infused at Mawson and was accompanied by an increase in pulmonary ventilation. However, the degree of the change was slight, and its significance is obscure because in rats the calorigenic effect of noradrenaline is concerned with nonshivering thermogenesis, which does not appear to play any important part in acclimatization as observed at Mawson.

The effects of short-term exposure to fairly severe wet cold conditions were investigated during the ANARE 1963 expedition to Heard Island [*Budd,* 1964*b, c*]. During their 6 weeks on the island the six members of the party worked outdoors in cold, wet, and windy weather, traveling extensively over rough terrain, backpacking all their gear in loads of 20–35 kg, and sleeping in unheated tents, huts, or emergency shelters. Their clothing was frequently wet from snow, rain, or immersion, and they often felt cold. Three of the men were exposed to additional cold stress during and after an attempt to climb Big Ben, the 2745-meter mountain that dominates the island. During the 16 days spent on the mountain they worked in almost continuous blizzard and air temperatures as low as −13°C. The loss of their tent, sleeping bags, spare clothing, and other equipment then forced an emergency descent in wet clothing from 2285 meters to sea level, during which they became severely chilled and one man's hands were frostbitten. For the next 9 days, until they had walked around the island to rejoin the rest of the party, they lived in makeshift shelters and slept in down suits, which remained sodden for several days and even when dry were inadequate for comfortable sleep.

Before and after the expedition (series 1 and 2, respectively) the six men underwent a test exposure in the cold room of the expedition ship [*Budd,* 1965], during which they sat in their underwear for 2 hours in an air temperature of 4°C, this exposure causing rectal temperature to fall by as much as 1.2°C. Body temperatures and oxygen consumption were measured in the same

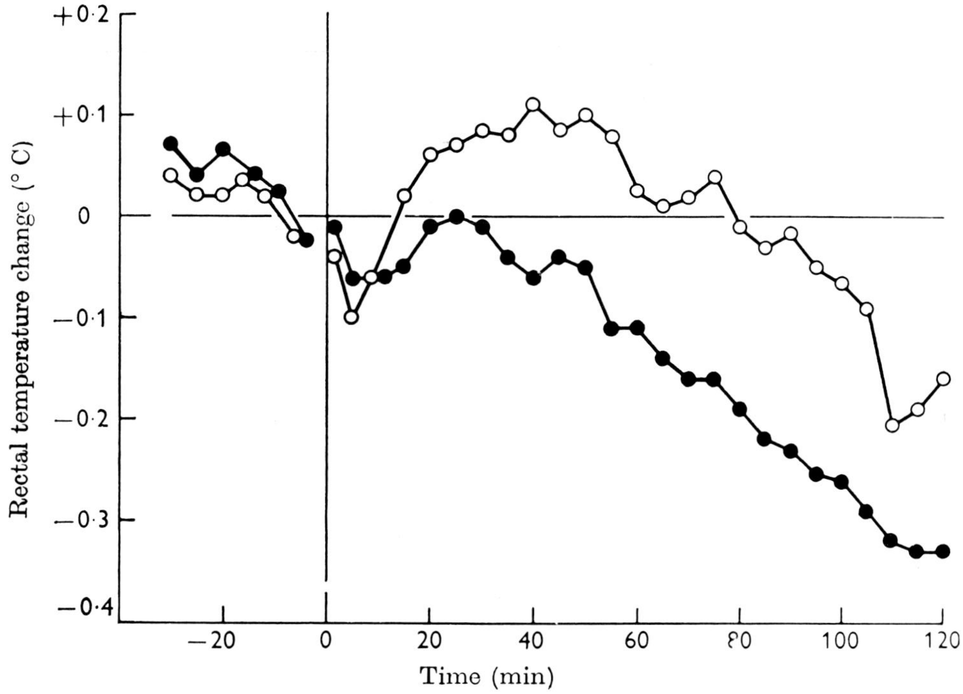

Fig. 10. Rectal temperature change in series 1 (filled circles) and series 2 (open circles). Mean values are given for the eight exposures in each series. Zero time indicates the commencement of the cold phase [*Budd and Warhaft,* 1966*a*].

way as they were at Mawson, but Scholander analysis of expired air was impractical in the ship, and so the samples were sealed in glass ampuls for later analysis in Australia. Unfortunately, more than half of them leaked, and the samples were lost. The subjects carefully noted the time of the onset of shivering and recorded their shivering intensity and thermal discomfort every 5 min in terms of standard 5-point scales. In each series, body weight, skinfold thickness, basal blood pressure, and basal heart rate were measured, and physical fitness was assessed by the Harvard step test.

The results show a significant increase in physical fitness and a decrease in skinfold thickness; the latter was correlated with weight loss (r = 0.823, $P < 0.05$), but the former was not. In series 2, rectal temperature tended to fall less rapidly than in series 1, and shivering was generally delayed in onset and was less intense. In the only two within-subject comparisons available, oxygen consumption in the test cold exposures was less in series 2 than in series 1. These findings appear to show that the reduction in body insulation caused by the loss of subcutaneous fat was balanced or even outweighed by an increase in the insulation of other tissues. Finger temperature fell more rapidly in series 2, there was less cold-induced vasodilatation, and the gradient of skin temperature between elbow and finger increased significantly, these findings suggesting that heat was conserved by means of countercurrent heat exchanges and enhanced vasoconstriction. The discomfort felt by the subjects during the test cold exposure did not change between series, even though an apparent habituation to cold had been noticed after the descent from the mountain, when shivering had gradually lost its power to interrupt conversation or prevent sleep.

The results of this investigation contrast with those of a study of Norwegian students living outdoors under somewhat similar conditions [*Scholander et al.,* 1958], who had higher skin temperatures and slightly higher metabolic rates than control subjects. They also contrast with previous observations showing that higher skin temperatures accompanied an increase in physical fitness [*Adams and Heberling,* 1958; *Heberling and Adams,* 1961; *Keatinge,* 1961]. On the other hand, they agree with the Mawson studies of 1959 and 1964 in suggesting that tissue insulation increased as a result of vasomotor changes.

At Little America 5 in 1957 [*Milan et al.,* 1961], eight men were exposed nude for 2 hours to air at 17°C in the autumn, winter, and spring. The main finding was that the metabolic response to the cold exposure was significantly less in winter and spring than in autumn, although skin temperatures were higher. Basal metabolic rate

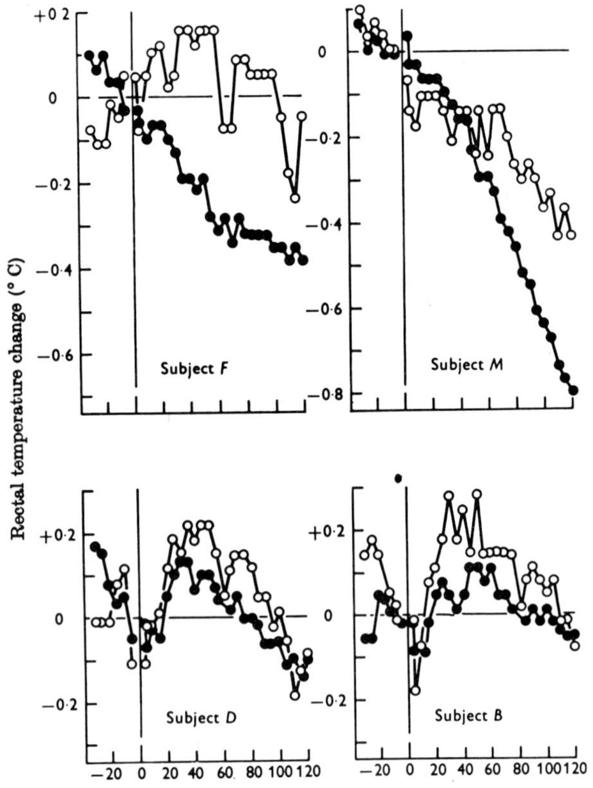

Fig. 11. Rectal temperature change in series 1 (filled circles) and series 2 (open circles). Individual values (mean of two replicates) are given in each series. Zero time indicates the commencement of the cold phase [*Budd and Warhaft,* 1966*a*].

remained constant. Rectal temperature was unaffected by the exposure, probably because the cold stress was too mild to provoke a measurable change. Body weight increased but in a different pattern from the decrease in metabolic rate: between autumn and winter, weight increased by 1.8 kg, and metabolic rate in the cold decreased by 12 kcal/m^2/hr, but a further weight gain of 0.6 kg between winter and spring was accompanied not by a further fall in metabolic rate but by an increase of 5 kcal/m^2/hr.

At Sanae [*Wyndham and Plotkin,* 1963; *Wyndham et al.,* 1964], six men were exposed nude for 1½ hours to air temperatures of 5°, 10°, 15°, and 27°C in the winter and summer of 1961. Their thermal and metabolic responses after 1 hour were compared with their own responses in South Africa the previous winter and also with the responses of four groups of 11 men previously tested, one group at each air temperature, in South Africa. The main finding was that oxygen consumption at the lower air temperatures was less in Antarctica than in South Africa. The authors concluded that tissue insulation had increased in Antarctica and attributed it to weight gain. The results of a later investigation [*Wyndham and Loots,* 1969] were thought to support this view and also to suggest that vasoconstrictor tone had increased. More recently, an enhanced ability to maintain rectal temperature has been reported at Casey [*Bodey,* 1973].

In summary, the subjects of the investigations described experienced their customary cold exposure under markedly differing circumstances. In four of the investigations the men lived for a year in compact stations, in two they lived for a year in an open station, and in one they spent 6 weeks on a wet and windy island. The test cold exposures, as well as the measurement techniques used in them, were equally varied. In spite of this diversity, certain broad similarities are apparent in the results. First, neither the basal metabolic rate nor the metabolic response to cold was found to increase, showing that man in the Antarctic does not resemble the cold-acclimatized laboratory rat. Second, all showed changes of one kind or another in the response to cold, which in each investigation were consistent with the hypothesis that tissue insulation had increased. At Sanae the increase was attributed mainly to fat deposition, but at Mawson and Heard Island and possibly at Little America 5 it appeared to be the result of an enhanced vasomotor response to cold.

Indirect evidence of general acclimatization. Frazier [1945] and *Butson* [1949], at Little America 3 and Stonington Island, respectively, formed the impression that men did not wear additional clothing as the year progressed and the weather became colder yet still felt comfortable, presumably because of acclimatization. *Goldsmith* [1960] undertook to obtain quantitative evidence of this phenomenon at Shackleton in 1956. Every week he recorded the clothing that men wore on the trunk and extremities when they were working outdoors, together with their 'comfort vote,' as described above. He found that, although they did wear extra clothing in winter, after winter they wore less than they had in periods of equal wind chill before it yet were no less comfortable. The difference in the amount of clothing worn was statistically significant ($P < 0.05$) for gloves and approached significance ($P < 0.10$) for footwear. However, since all garments appeared to have lost insulation from soiling and wear, it seems likely that on the trunk too the men were keeping warm after winter with less insulation than before winter. That the evidence appears to be strongest for the hands is consistent with previous demonstrations [*Mackworth,* 1953, 1955; *Massey,* 1959] of acclimatization of the hands and with the observation that the hands were more frequently cold than the trunk, thus sustaining a greater stimulus to acclimatization.

Palmai [1962*a*] repeated the investigation at Macquarie Island in 1960 and obtained very similar results. Clothing increased in winter, and after midwinter the men wore less clothing on the trunk than they had worn in the months of equal wind chill before it. Not only was the reduction in trunk clothing statistically significant, but the comfort vote was also significantly warmer. The same result was obtained in a further, more rigorous comparison, in which Palmai compared days of observation before and after midwinter that were equivalent not only in wind chill but also in the subjects' activity and in the elevation of the sun, for they were equidistant from the winter solstice. In contrast to the observations at Shackleton, the clothing worn on the hands and feet changed little, probably because, as Palmai suggested, gloves were worn mainly for mechanical protection, and the leather boots worn did not permit much variation in the number of socks.

Lugg [1965] made a similar investigation at Davis in 1963. He found that, although the number of pairs of gloves worn increased in the winter, the amount of clothing worn on the head, trunk, and feet decreased in each successive quarter of the year. Part of the decrease seems due to accidents of sampling, for it so happened that the wind chill at the time of the weekly observations was practically the same in the winter quarter as it was in the preceding and following quarters, even though winter was undeniably colder than the other seasons and was no less windy. Despite the decrease in the clothing worn, the comfort votes were warmer in each successive

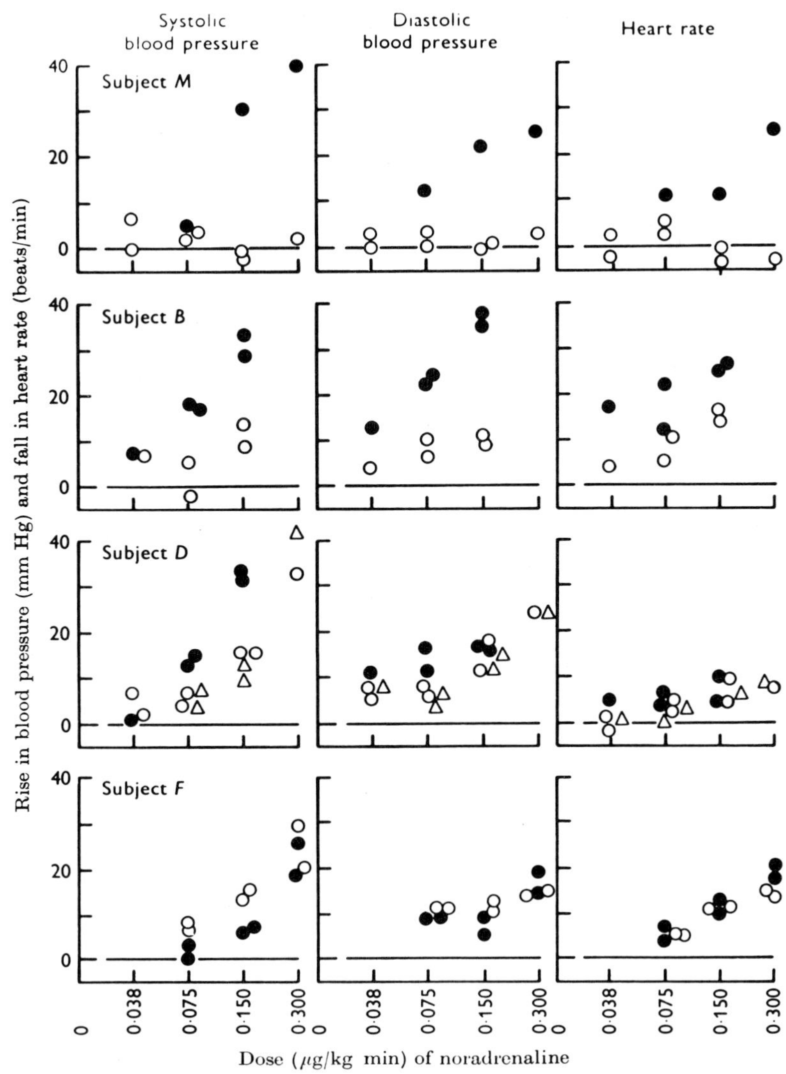

Fig. 12. Response of blood pressure and heart rate to various doses (plotted on a logarithmic scale) of noradrenaline in Melbourne (filled circles, series 1) and Mawson (open circles, series 2; triangles, series 3). Each point represents one observation [*Budd and Warhaft,* 1966*b*].

quarter. To test for acclimatization, Lugg compared the means of observations made in the 17 weeks immediately before midwinter with those made in the 17 weeks immediately after it. Although the mean values were not tested statistically, they appeared to show that the two periods were equivalent in wind chill, sunshine, and activity and that after midwinter the men felt warmer, although they were wearing less clothing, than they had felt before midwinter.

These investigations provide suggestive evidence of acclimatization. The main alternative explanation, that the men might have overdressed at first (as newcomers to cold climates usually do) and later reduced their clothing to the amount that was really necessary, seems unlikely because a number of members of each expedition had wintered in polar regions before and because the comparisons made have usually included periods some months after the party's arrival, by which time the inexperienced men would have learned how much clothing to wear. *Palmai*'s [1962*a*] observations are of particular interest because they provide the strongest evidence of general acclimatization so far obtained by this method and thus emphasize the possibility that the wet cold subantarctic climate, in which it is difficult to keep clothing dry, might impose a greater stimulus to acclimatization than the dry cold climate of the antarctic continent.

It is worthwhile to consider how these changes in clothing and thermal comfort might have come about. A decrease in clothing insulation in a constant environment would seem to imply either increased heat production to balance the presumably increased heat loss or else a compensatory increase in the insulation of the tissues.

Basal metabolic rate has been measured on many occasions in the Arctic and Antarctic, and the results, summarized by *Wilson* [1965], have shown either no seasonal change or a decrease in winter. K. E. Hicks (personal communication, 1968) obtained similar results at Wilkes in 1965 when he measured the basal oxygen consumption of 17 men over the 6 months March to August and found that the monthly means varied between 270 ml/min and 277 ml/min. The metabolic response to acute cold stress, described in the previous section, has shown either no change or else a decrease with acclimatization; and there is no evidence to suggest that the energy cost of specific tasks increases [*Edholm and Lewis,* 1964]. It therefore seems unlikely that the changes in clothing and comfort could be due to changes in heat production.

A ready explanation of increased tissue insulation might lie in the increased fatness that so many observers (summarized by *Wilson* [1965]) have reported in polar regions, but it does not suffice in the present instance. *Goldsmith*'s [1959] subjects did not gain weight; *Palmai*'s [1962*a*] subjects were about 1 kg heavier in the observations after midwinter than they had been in those before midwinter, but their skinfold thickness was less or unchanged more often than it was greater; and Lugg's subjects showed negligible changes in body weight and skinfold thickness throughout the year. If increased tissue insulation was the reason for the changes in clothing and thermal comfort, it was evidently not brought about by fat deposition and presumably must have been the result of vascular changes.

Decreased clothing insulation and increased tissue insulation alike would seem to imply lower skin temperatures and hence a change in the relation between thermal comfort and skin temperature. *Adam* [1959] recorded the comfort vote and subclothing temperature of members of the Commonwealth Trans-Antarctic Expedition who were nearing the end of their journey and compared the results with similar observations on men newly arrived in the Antarctic. The results, he suggested, did not show any differences between the groups. On the other hand, observations on sedentary men indoors at Mawson in 1959 suggest that skin temperature in each grade of thermal comfort was lower after midwinter than before it. However, the results are not conclusive, and they are not supported by the measurements of skin temperature made when men were resting under blankets immediately before the test cold exposures [*Budd,* 1964*a*].

Acclimatization of the hands. It was noted in the section on thermal stress in the Antarctic that the hands were uncomfortably cold in 17% of more than 2000 observations made on men working outdoors and in as many as 33% of the observations made in the coldest weather. The main reasons for this high incidence of discomfort are that cold hands often accompany a cold trunk because of general vasoconstriction, that even the best mittens or gloves are of limited insulation [*van Dilla et al.,* 1949], and that when manual dexterity is required or when wetting is likely men usually have to work with bare hands. As well as being more frequent, the cold stress sustained by the hands is often more severe than that sustained by other parts of the body, for they are often painfully cold and occasionally become frostbitten. There would thus appear to be ample stimulus for the development of acclimatization.

Mackworth [1953] observed that the hands of men who habitually worked outdoors in the Arctic were more resistant to the numbing effects of cold than those of indoor workers, and he later managed to induce such resistance by means of cold exposures in the laboratory [*Mackworth,* 1955]. *Massey* [1959], using the same technique at Hope Bay in the Antarctic, found that men's fingers became more resistant to numbing within 6 weeks after their arrival and changed little thereafter. This adaptation might have resulted either from improved nerve conduction at low finger temperatures or from an increased blood flow that kept the fingers warm during cold exposure. The latter possibility was favored by investigations of fish filleters, Eskimos, and other groups suffering chronic cold exposure, for their hands when immersed in ice water showed either a greater blood flow than those of control groups [*Brown et al.*, 1954] or an earlier onset and greater magnitude of rewarming by cold-induced vasodilatation [*Yoshimura and Iida*, 1952; *Elsner et al.*, 1960; *Nelms and Soper*, 1962].

Elkington [1968] therefore made observations of finger blood flow before and during a 1-hour immersion in ice water on 25 members of the 1966 Wilkes party. He used a strain gage plethysmograph and made a total of 121 satisfactory immersions. The results were contrary to expectation. Finger blood flow before and during immersion, instead of increasing, decreased in the colder months of the year, and the increase caused by cold-induced vasodilatation became less as the year progressed. The time of onset, blood flow at onset, and frequency of cycles, of vasodilatation showed no significant relation to the mean monthly wind chill or the time in months. Similar results were obtained after the more severe cold exposure sustained in five inland journeys of an average duration of 6 weeks.

These results suggest that vasoconstrictor tone increased, a conclusion that agrees well with *Massey*'s [1959] finding that finger skin temperature was lower after the coldest part of the year. Calorimetric observations by *Hampton* [1969] at Hope Bay, described by *Edholm* [1974], have also led to the same conclusion. The contrast with other investigations showing increased finger blood flow was attributed by Elkington to either of two causes. The first was that the Wilkes investigation was a longitudinal study on a single group of men, whereas the other investigations compared two or more groups, who may have differed in other respects (such as race, diet, and physical fitness) besides cold exposure. The second was that the changes in finger blood flow at Wilkes may have been part of a process of general acclimatization rather than of local acclimatization of the hands alone. The latter interpretation agrees well with the results of work on general acclimatization at Mawson and Heard Island and serves to emphasize the somewhat arbitrary nature of the distinction between local acclimatization and general acclimatization.

Acclimatization of the face. The face is probably exposed to greater cold stress than any other part of the human body. No satisfactory form of clothing has yet been devised for it, because masks accumulate ice from the breath and the nasal secretions and become so inconvenient that most men prefer to leave the face bare. Although some degree of shelter from wind can be obtained by adjustments of a parka hood, the face is nevertheless often frostbitten in cold winds. Thus there would appear to be a strong stimulus to adaptations conferring increased comfort and resistance to frostbite.

Resistance to frostbite was described by *Frazier* [1945], who exposed men to cold winds at Little America 3 and reported that the faces of men habitually outdoors ('outdoor men') took much longer to become frostbitten than did those of men habitually indoors ('indoor men'). Unfortunately, Frazier did not publish the evidence on which this conclusion was based, and, since the conclusion is an important one, repetition of the experiment seemed desirable.

Adaptations conferring increased comfort are suggested by subjective impressions. Exposure of the face to a wind of 15 m/sec at −8°C on arrival at Mawson in late summer caused characteristic and unpleasant symptoms. The face became painfully cold and then numb, the eyes watered, the nose smarted and ran, and the facial muscles became stiff, making it difficult to speak clearly. Within a few weeks this triad of pain, mucosal irritation, and impairment of sensory and motor function had diminished, although the weather was no warmer. Falling temperatures with the approach of winter again induced it but only temporarily, so that, although cold stress was steadily increasing, facial discomfort was not. In late winter, whenever the temperature rose after a cold spell the wind felt mild and pleasant, even though it was stronger and colder than winds that earlier in the year had caused intense facial discomfort.

To test these impressions and to investigate their physiological basis, three pairs of experiments were done at Mawson in 1959 and 1960 and at Melbourne in 1962 (G. M. Budd, unpublished observations). Following a 1-hour control period indoors the faces of two indoor men and two outdoor men were simultaneously exposed to cold winds for 5 min, and the skin temperature and thermal comfort of their faces were observed every 2 min during the next hour's rewarming indoors. Further experiments done when the relief ship arrived in 1960 compared newcomers with men who had spent the year at Mawson, and 2 years later the original subjects were retested in a Melbourne cold room that duplicated the environmental conditions of the 1959 exposures. No frostbites occurred in any experiment. The results for skin temperature and thermal comfort, although they are inconclusive because of the unavoidable smallness of the groups and the presence of large differences between the subjects, suggest that cold-conditioned men become habituated to low facial temperature. This was particularly apparent in two subjects, whose faces, tested 2 days after their return from a 12-day sledging journey in the spring, felt warmer than they had at the same skin temperatures before the journey.

A further series of experiments was done by J. W. Hudson (personal communication, 1967) at Mawson in 1966. At approximately monthly intervals throughout the year the facial responses to cold of 10 outdoor men and three indoor men were observed. These groups included most of the outdoor men in the expedition but only a small proportion of the indoor men, the remainder being reluctant to undergo the test cold exposures, a difference that itself might indicate lesser cold tolerance in indoor than in outdoor men. Control observations were made during a 1-hour preliminary period in standardized conditions indoors, and the 13 subjects then went outside and stood facing the wind for 10 min. Air temperature during the year's exposures ranged from -28° to 5°C, wind speed (at the meteorological hut 100 meters away) ranged from 0 to 22 m/sec, and wind chill ranged from approximately 600 to 1900 units. At the end of the exposure the men returned indoors and recorded their facial and other sensations in terms of standard categories.

The members of both groups wore their normal outdoor clothing yet felt too cold on the clothed body in 38% of the 151 exposures. Shivering occurred in 26% of the indoor men's exposures and in 16% of the outdoor men's. The men's faces felt too cold in 46% of the tests and were numb in 19%; nasal drip was reported in 43%, and watering eyes were reported in 15%.

It was expected that the outdoor men, as a result of either self-selection or acquired acclimatization, would tolerate the exposure better than the indoor group. The results did not support this expectation. The outdoor men's faces felt uncomfortably cold just as often as those of the indoor men, their noses ran just as often, and facial numbness and watering eyes occurred twice as often as they did in the indoor men. Perhaps because the experiments were done on days when there was no drifting snow (which increases cold stress to an unknown extent), frostbite occurred only 3 times in 151 tests, each time in an outdoor man. Six of the outdoor men took part in a 4-month journey inland that involved prolonged cold exposure, but when they were tested at Mawson after their return their faces felt just as cold as those of the men who had stayed behind. Striking changes occurred from month to month in all subjects, but unfortunately it proved impossible to compare occasions of equal cold stress (thus testing the original subjective observations) because of a fault in the anemometer used to measure wind speed at the exposure site.

Two factors might serve to explain the difference between the results of this experiment and those of *Frazier*'s [1945] experiment. First, the test cold exposures were less severe than his and sufficed to frostbite only a few men in the 10-min exposure period. Second, the three indoor men who volunteered might have been the most cold tolerant or most cold exposed of all the indoor men at the station and hence would have been an unsuitable control group for the outdoor men. Further experiments, in which larger groups of men are exposed for longer periods (if possible to a standardized cold stress) and additional variables such as skin temperature are measured, would seem desirable, in order to see what adaptations do occur in this most cold-stressed part of the human body.

Summary of results on acclimatization. Direct evidence of general acclimatization has been obtained twice at Mawson and Sanae and also at Little America 5, Heard Island, and Casey. Indirect evidence has come from Shackleton, Macquarie Island, and Davis. Direct evidence of local acclimatization has been obtained at Wilkes and twice at Hope Bay, and indirect evidence has been obtained at Shackleton. Changes in blood coagulation similar to those seen in cold-adapted animals (described in the section on blood coagulation studies) have been observed twice at Wilkes and also at Mawson and Macquarie Island.

As regards the stimulus to acclimatization, the enhanced vasoconstriction that mediated acclimatization at Mawson appears to have been a remarkably appropriate adaptation to the alternating thermal stress that the subjects experienced at the station and in the field. Whether the stress experienced in the field or that experienced at the station was the more important is not known, for the subjects were all men who worked in both places. However, the changes in thermal comfort and clothing at Macquarie Island, which occurred in all members of the party, suggest that at an open station indoor men too can acclimatize. This might be true even of compact stations, for *Milan et al.* [1961] found no difference between the responses of indoor men and those of outdoor men at Little America 5, and *Elkington* [1968] found that the changes in finger blood flow that developed in the whole party throughout the year were of the same kind as those occurring during field trips.

These considerations emphasize the fact that as yet little is known about human adaptation in the Antarctic. The work done so far has been largely exploratory in nature and has yielded the important conclusion that man does acclimatize to the cold. Why and how he does so are the two main questions awaiting future work.

Future work. The first question, What is the optimum stimulus to acclimatization?, might be approached in two ways. First, parallel studies by means of whole-body cooling at various stations could provide comparisons to assess the influence of station design (Mawson versus Casey), of climate (the dry cold of Mawson versus the wet cold of Macquarie Island), and of occupation (indoor men versus outdoor men). Second, concurrent studies of the patterns of thermal stress to which the same subjects would be exposed could be made by assessing the average levels and variability of their thermal comfort, skin temperature, activity, clothing, and in some cases their deep body temperature. Correlation of these observations with the subjects' responses to cooling might prove instructive. An important practical application of this work would be that any conclusions drawn about the optimum stimulus to acclimatization could expedite the development of one of the most necessary tools for further progress in this field, a reliable means of acclimatizing men to cold in the laboratory.

The second question, What is the mechanism of acclimatization?, can best be answered when it is known what occupational groups at what stations develop the highest degree of acclimatization. The conclusions drawn so far can be examined further by such methods as bath immersion studies of tissue insulation, measurements of body fat, and plethysmographic determinations of the vascular response to whole-body cooling. Further work on the excretion of adrenal hormones and their metabolites and on the mechanism of the changes observed in blood coagulation (described in the section on blood coagulation studies) is already in progress. One of the most interesting investigations, in view of the alternating thermal stress experienced by men in the Antarctic, will be the determination of their adaptation to heat as well as to cold, for it may be that their adaptation to both will improve together.

BODY WEIGHT, SKINFOLD THICKNESS, BLOOD PRESSURE, HEART RATE, AND ORAL TEMPERATURE

From the time of the earliest antarctic expeditions it has been reported that men put on weight in the winter, presumably from the deposition of extra fat. This 'adiposea of the south,' as it used to be called, has generally been assumed to be an insulative adaptation to winter cold. However, observers on two modern expeditions that provided evidence of the weight gain [*Lewis et al.*, 1960; *Wilson,* 1960] did not endorse this view. Pointing out that the expedition members tended to stay indoors during winter, they suggested that the weight gain resulted not from extremes of cold exposure but from the more prosaic experience of overeating during a period of restricted activity. Lewis et al. also measured skinfold thickness and found that it varied in a manner similar to the way that body weight varied, thus supporting the view that the weight gain was due to fat. *Palmai* [1962*b*], however, repeating this work at Macquarie Island in 1960, found that the monthly variations in weight and skinfold thickness were not correlated and suggested that the relationship may have been obscured by the effects of muscular hypertrophy at times of strenuous exertion. Palmai also measured blood pressure and found that its monthly changes were highly correlated with those of skinfold thickness (but not body weight), an interesting finding with apparent implications for the clinical management of obesity and hypertension.

Since 1962 the medical officers at all ANARE stations have made routine monthly measurements by standardized methods of body weight, skinfold thickness, blood pressure, heart rate, and oral temperature under both basal conditions (early morning before rising) and casual conditions (afternoon), and several thousand observations have been accumulated. An example of the results from one year is given by *Hicks* [1966].

Simple routine observations of this kind have a number of applications. First, they provide an adequate sample for testing some of the generalizations of polar folklore. As Figure 13 shows, weight increase in winter occurs often enough to be suggestive, but it is by no means invariable, and its pattern may vary between stations, years, and occupational groups, such as the men taking part in the main field trips ('field group') and those remaining behind ('base group'). Possible reasons for these variations may be sought in differences of weather, station design, diet, activity, and many other factors. Regression analysis on such variables as air temperature, hours of daylight, and wind chill can prove useful in evaluating the possible influence of the environment on the physiological changes observed, as it did in a recent investigation of finger blood flow at Wilkes [*Elkington,* 1968]. But even if no satisfactory explanation can be found for the diversity of the changes observed, the mere recognition of this diversity is itself of importance and emphasizes the need for caution in interpretation.

More importantly, the observations in each station and year are in effect a multivariate longitudinal study of a population of healthy men, usually young, who spend a year or more in a simple way of life, the main features of

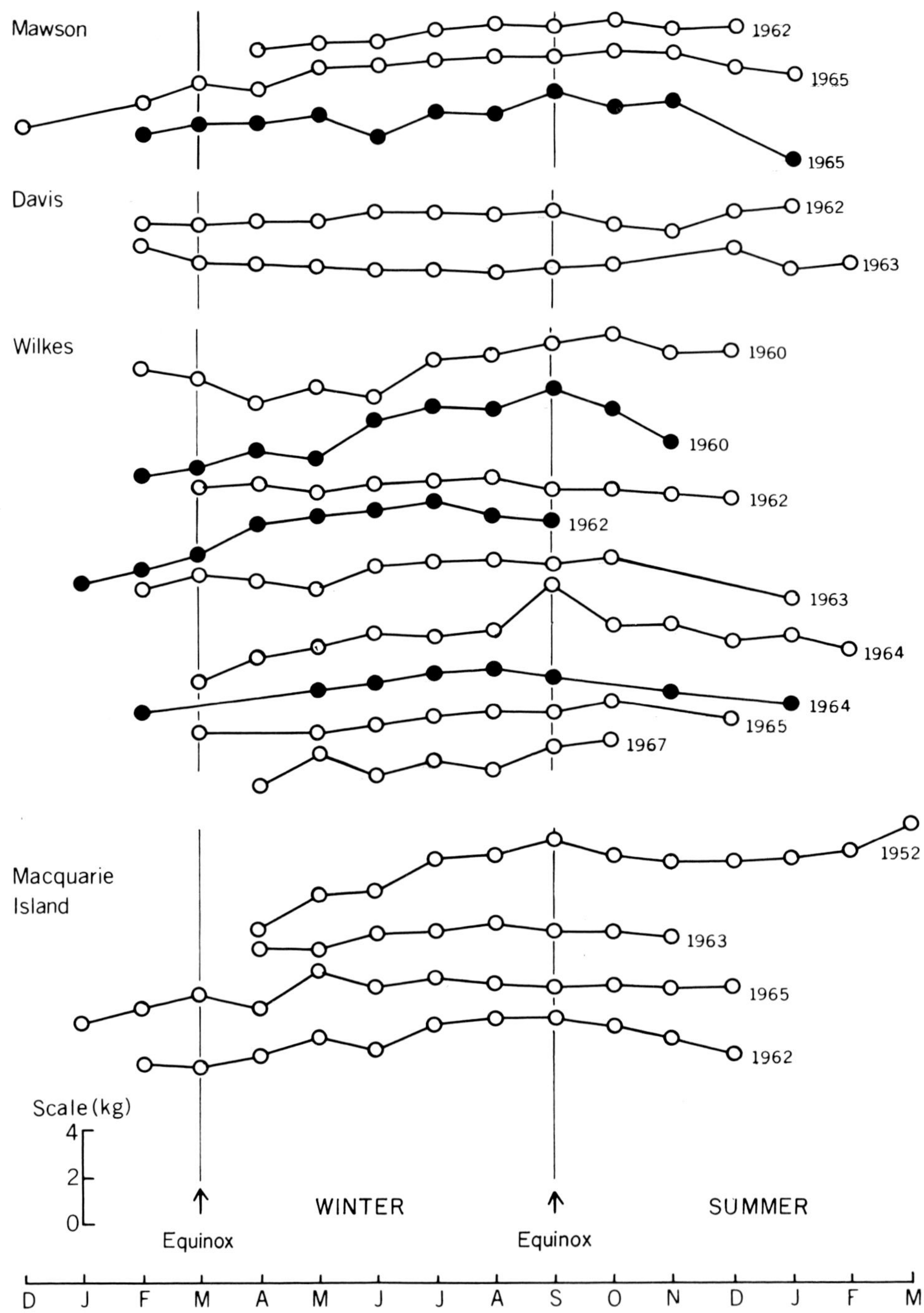

Fig. 13. Monthly variations in body weight at ANARE stations in various years. Means for field (filled circles) and base (open circles) groups were computed separately when the field group missed a month or more of the observations.

which can readily be described or measured by the single observer who shares it with them. Longitudinal observations of this kind, especially basal observations made on waking, are difficult to make elsewhere without hospitalizing the subjects, and they provide an opportunity for furthering our knowledge of physiological processes that in themselves may have little to do with the antarctic environment. In this way, polar expeditions might be considered as 'human laboratories,' which provide opportunities for the study of fundamental human biology [*Goldsmith and Lewis,* 1960].

A simple example is provided by observations of resting heart rate. *Durnin* [1965] has suggested that the resting heart rates of healthy young adults, measured in bed after waking but before rising in the morning, may be lower than is generally thought. He found that in a group of 44 young men the average was about 55 beats/min. The basal observations made at ANARE stations support this view, for the yearly mean values from eight parties ranged from 56 to 63 beats/min, and the average was 60 beats/min.

The interrelations among the different variables measured may be examined by a variety of statistical techniques [*Cattell,* 1966], of which multiple analysis of covariance has been used for the preliminary examination of the ANARE results. Analysis of covariance permits three sets of linear correlations and regressions to be calculated and tested for significance: those associated with differences between subjects, those associated with differences between months, and the residual, or 'error,' variation remaining after the variation associated with subjects and months has been accounted for. Two examples illustrate the method.

Body weight and skinfold thickness. Table 6 shows the correlation between body weight and skinfold thickness at four stations over a number of years. Considerable variability is apparent in the subject and month terms. The subject term shows that, although in some years the heavier men tended to be the fatter ones, in most years weight and skinfold thickness were not significantly correlated, possibly because of differences in body build. The month term shows that the monthly fluctuations in the average weight of the party were usually not paralleled by those of skinfold thickness, a finding in keeping with the observations of other workers [*Massey,* 1956; *Palmai,* 1962*b*; *Orr,* 1965; *Hicks,* 1966; *Easty,* 1967] who have reported a dissociation between monthly changes of weight and those of skinfold thickness in the Antarctic. In the error term, significant correlations were found in almost every year, but they were generally weaker than those for the subject or month terms. That most of them are significant merely reflects the fact that, with enough degrees of freedom, even small values of the correlation coefficient r may be significantly different from zero. The error correlation coefficients themselves were mostly below 0.5 and thus indicate that less than 25% (r^2) of the variation in body weight was attributable to its correlation with skinfold thickness [*Snedecor and Cochran,* 1967].

Blood pressure, body weight, and arm circumference. Many surveys have shown that arterial blood pressure, as measured by a sphygmomanometer, is correlated with body weight [*Lowe,* 1964]. This, however, is a between-subject correlation showing the extent to which heavier people tend to have higher blood pressures; it does not strictly imply that weight reduction by a group of people would be likely to result in a lowering of their blood pressures. Only longitudinal studies can yield this kind of prediction. Assessment of the relation between blood pressure and body weight has been complicated further by the finding that both are correlated with arm circumference [*Ragan and Bordley,* 1941; *Khosla and Lowe,* 1965]. Some authors [*Pickering et al.,* 1954] have therefore 'corrected' their blood pressure readings for arm circumference, but others [*Lowe,* 1964; *Khosla and Lowe,* 1965] have concluded that the correlation between arm circumference and blood pressure merely reflects the correlation of both with body weight and that correcting for arm circumference would obscure the important influence of body weight.

The ANARE results have consistently shown significant correlations between body weight and arm circumference in both the subject term and the error term, showing that not only do heavier men tend to have larger arms (as was observed previously) but that changes in body weight tend to be accompanied by corresponding changes in arm circumference. Equally consistent error correlations between skinfold thickness and arm circumference suggest that subcutaneous fat might be the common factor affecting both weight and arm circumference, but the partial correlations make this seem doubtful. Holding skinfold thickness constant had little effect on the correlation between arm circumference and weight, whereas holding weight constant removed the correlation between arm circumference and skinfold thickness. Evidently the changes in arm circumference were associated with some factor that affected body weight but not skinfold thickness, this finding suggesting that variations in muscularity rather than fat may have been responsible. Blood pressure, however, has only infrequently been correlated with arm circumference, body weight, or skinfold thickness in the month, subject, or error term. The ANARE results as a whole therefore provide little evidence to suggest that reduction in weight or fatness would be accompanied by a lowering of blood pressure.

BLOOD COAGULATION STUDIES

Many accounts of life in the Antarctic have mentioned that men ate more fatty foods than they did in temper-

TABLE 6. Coefficients of Correlation of Body Weight with Skinfold Thickness of the Triceps (Arm) and Subscapular (Back) Regions, Together with Their Degrees of Freedom, in the Subject, Month, and Error Terms of the Analysis of Covariance

Station	Year	Observer	Field or Base	Subjects			Months			Error		
				d.f.	Arm	Back	d.f.	Arm	Back	d.f.	Arm	Back
Mawson	1962	D. R. Wigg	Both	8	...	0.352	7	...	−0.109	71	...	0.649§
Mawson	1965	A. S. Cameron	Base	16	−0.081	0.163	11	0.359	0.757‡	203	0.638§	0.656§
Mawson	1965	A. S. Cameron	Field	7	−0.189	0.335	9	0.006	−0.094	73	0.440§	0.547§
Davis	1962	J. Boda	Both	7	0.502	0.278	10	0.171	0.049	87	0.490§	0.487§
Davis	1963	D. J. Lugg	Both	6	−0.311	−0.099	11	−0.566*	−0.188	76	0.308‡	0.290‡
Wilkes	1962	Z. Soucek	Base	14	0.080	0.156	8	0.218	0.721†	134	0.172†	0.326§
Wilkes	1962	Z. Soucek	Field	4	0.788*	0.138	7	0.618*	0.526	39	−0.029	0.203§
Wilkes	1963	K. E. Hicks	Both	16	−0.015	...	8	0.444	...	152	0.479§	...
Wilkes	1964	L. G. Murray	Base	11	0.459	0.535*	10	0.138	0.278	131	0.304§	0.353§
Wilkes	1964	L. G. Murray	Field	8	0.521	0.577*	6	0.753†	0.288	62	0.524§	0.411§
Wilkes	1965	K. E. Hicks	Both	21	0.680§	0.755§	6	0.490	0.184	153	0.228‡	0.224‡
Wilkes	1967	R. I. Liddell	Both	10	0.776‡	0.718‡	5	0.603	0.677*	65	0.008	0.480§
Macquarie Island	1963	L. G. Murray	Both	14	0.473*	0.689‡	6	0.904‡	0.679*	104	0.236†	0.155
Macquarie Island	1965	H. K. Lim	Both	13	0.668‡	0.722‡	10	0.183	−0.167	153	0.361§	0.466§
Macquarie Island	1968	J. Boda	Both	15	0.078	0.319	9	0.571*	0.588*	159	0.263§	0.470§

Whenever the men who went on major field trips (the field group) missed several months of observations, their results for the year were analyzed separately from those of the men who remained at the station (the base group). Otherwise, the results from both groups were pooled for analysis.

Coefficients without superscripts are not significantly different from zero.

*$0.10 \geqslant P > 0.05$.

†$0.05 \geqslant P > 0.01$.

‡$0.01 \geqslant P > 0.001$.

§$0.001 \geqslant P$.

ate climates, and that they suffered emotional stress from the isolation as well as physical stress from the cold. K. E. Hicks thought it likely that these factors would produce a rise in serum cholesterol and a fall in the blood-clotting time. He therefore supplemented his routine monthly observations of body weight, blood pressure, etc., in the 18 members of the 1963 Wilkes party with determinations of serum cholesterol, blood-clotting time, and prothrombin index [*Hicks,* 1965]. Serum cholesterol showed the expected rise, but clotting time, far from falling, showed a marked increase in late winter and spring, returning to normal levels by the end of the year. In 1965 Hicks returned to Wilkes and repeated the investigation: he conducted tests for various blood coagulation factors and made measurements of body density by water displacement, of total body water by the deuterium dilution method, and of basal metabolic rate. Again, clotting time increased [*Hicks,* 1967], but, on this occasion, serum cholesterol did not except in the first month. Hicks suggested that the increase in clotting time was caused by a decrease in plasma thromboplastic lipid.

Two possible explanations for the changes in blood coagulation were proposed. One is that they reflect reduced emotional stress in a harmonious party, and this view is supported by a significant negative correlation between the monthly means of clotting time and those of basal blood pressure. The other possibility is that the changes in clotting time are an accompaniment of acclimatization to cold, for similar changes have been observed in cold-adapted rabbits and hamsters. Similar increases in blood-clotting time have since been observed at Macquarie Island by H. K. Lim (personal communication, 1966) and at Mawson by J. W. Hudson (personal communication, 1969), and further work is in progress.

The role of a high-fat diet, however, may be less than Hicks originally supposed. M. N. Orton tested the common belief that in the Antarctic men eat more fat and absorb it better. He carried out a series of 39 four-day fat balance studies at Wilkes and in Australia. He found, contrary to expectation (M. N. Orton, personal communication, 1962), that men ate and absorbed much the same amount of fat in both places. Moreover, seasonal and occupational differences in cold stress had no clear effect on either fat intake or its absorption.

SEASONAL AND DIURNAL VARIATIONS IN DEEP BODY TEMPERATURE

Basal oral temperature, measured once a month on all available men at ANARE stations and on many other expeditions (summarized by *Wilson* [1965]), has not been found to show any regular seasonal variation. However, *Palmai* [1962*c*] used a different method at Macquarie Island in which he measured every fortnight the oral temperature of just one subject (himself) at hourly intervals for 24 hours. He found that the form of the diurnal variation did not change with the seasons but that its level did, for the means of the 48 observations in each month varied with the outdoor air temperature, from 36.7°C in summer to 36.4°C in spring. In this variation of oral temperature with air temperature, Palmai's results resemble those of *Kleitman and Ramsaroop* [1948] in the United States, *Driver* [1958] in Hong Kong, and *Adam and Ferres* [1954] in Oxford and Singapore. Moreover, *Kleitman and Ramsaroop* [1948] found that oral temperature varied with the outdoor air temperature only in those months of the year when heating was not used indoors. Palmai's living quarters were unheated the year round, and this fact, together with the intensive technique that he used, may explain why his results differ from those of other observers in the Antarctic. Although further intensive investigations of this kind would be worthwhile, marked changes in diurnal rhythms are perhaps unlikely to be found at ANARE stations because, at their relatively low latitude (near 70°S), diurnal changes of daylight and darkness are apparent in all seasons of the year and because, as a matter of policy, most men follow a daily routine similar to what they follow in temperate climates.

Whether acclimatization to cold causes any change in deep body temperature measured in a warm environment is uncertain. At Mawson in 1964 the mean rectal temperature, measured immediately before the test cold exposures, was significantly lower (by 0.37°C) after acclimatization than before it [*Budd and Warhaft,* 1966*a*]. However, in the five series of tests in 1959 [*Budd,* 1964*a*] the mean preexposure rectal temperature varied over a range of only 0.21°C, and, although this variation was statistically significant, it showed no apparent relation to either air temperature or the degree of acclimatization.

Acknowledgments. The assistance of J. Robinson, Department of Mathematical Statistics, University of Sydney, and the willing cooperation of all the ANARE men who have been the subjects of the investigations, are gratefully acknowledged.

REFERENCES

Adam, J. M., Subjective sensations and subclothing temperatures in Antarctica, *J. Physiol. London, 145,* 26–27P, 1959.

Adam, J. M., and H. M. Ferres, Observations on oral and rectal temperatures in the humid tropics and in a temperate climate, *J. Physiol. London, 125,* 21P, 1954.

Adams, T., and B. G. Covino, Racial variations to a standardized cold stress, *J. Appl. Physiol., 12,* 9–12, 1958.

Adams, T., and E. J. Heberling, Human physiological responses to a standardized cold stress as modified by physical fitness, *J. Appl. Physiol., 13,* 226–230, 1958.

Amundsen, R., *The South Pole,* 1st ed., vol. 2, p. 194, Murray, London, 1912.

Aschoff, J., Kreislaufregulatorische Wirkungen der Kältedilatation einer Extremität als Folge extremer, umschriebener Abkühlung, *Pfluegers Arch., 248,* 436–442, 1944.

Bazett, H. C., L. Love, M. Newton, L. Eisenberg, R. Day, and R. Forster, Temperature changes in blood flowing in arteries and veins in man, *J. Appl. Physiol., 1,* 3–19, 1948.

Bedford, T., The warmth factor at work, *Rep. 76,* p. 18, Ind. Health Res. Board, London, 1936.

Belding, H. S., H. D. Russell, R. C. Darling, and G. E. Folk, Thermal responses and efficiency of sweating when men are dressed in arctic clothing and exposed to extreme cold, *Amer. J. Physiol., 149,* 204–222, 1947.

Bodey, A. S., The role of catecholamines in human acclimatization to cold: A study of 24 men at Casey, Antarctica, in *Polar Human Biology,* edited by O. G. Edholm and E. K. E. Gunderson, pp. 141–149, Heinemann Medical Books, London, 1973.

Brown, G. M., G. S. Bird, T. J. Boag, L. M. Boag, J. D. Delahaye, J. E. Green, J. D. Hatcher, and J. Page, The circulation in cold acclimatization, *Circulation, 9,* 813–822, 1954.

Budd, G. M., Acclimatization to cold in Antarctica as shown by rectal temperature response to a standard cold stress, *Nature London, 193,* 886, 1962.

Budd, G. M., General acclimatization to cold in men studied before, during and after a year in Antarctica, *ANARE Rep. 70,* 84 pp., Antarctic Div., Dep. of External Aff., Melbourne, 1964*a*.

Budd, G. M., The ANARE 1963 expedition to Heard Island, *ANARE Rep. 74,* 53 pp., Antarctic Div., Dep. of External Aff., Melbourne, 1964*b*.

Budd, G. M., Heard Island expedition, 1963, *Polar Rec., 12,* 193–195, 1964*c*.

Budd, G. M., Effects of cold exposure and exercise in a wet, cold antarctic climate, *J. Appl. Physiol., 20,* 417–422, 1965.

Budd, G. M., Skin temperature, thermal comfort, sweating, clothing and activity of men sledging in Antarctica, *J. Physiol. London, 186,* 201–215, 1966.

Budd, G. M., and N. Warhaft, Body temperature, shivering, blood pressure and heart rate during a standard cold stress in Australia and Antarctica, *J. Physiol. London, 186,* 216–232, 1966*a*.

Budd, G. M., and N. Warhaft, Cardiovascular and metabolic responses to noradrenaline in man, before and after acclimatization to cold in Antarctica, *J. Physiol. London, 186,* 233–242, 1966*b*.

Budd, G. M., and N. Warhaft, Urinary excretion of adrenal steroids, catecholamines and electrolytes in man, before and after acclimatization to cold in Antarctica, *J. Physiol. London, 210,* 799–806, 1970.

Budd, G. M., K. E. Hicks, D. J. Lugg, L. G. Murray, and D. R. Wigg, Thermal discomfort in the Antarctic and Subantarctic, *Med. J. Aust., 2,* 1285–1288, 1969.

Burton, A. C., and O. G. Edholm, *Man in a Cold Environment,* pp. 83–84, Arnold, London, 1955.

Burton, A. C., and R. M. Taylor, A study of the adjustment of peripheral vascular tone to the requirements of the regulation of body temperature, *Amer. J. Physiol., 129,* 565–577, 1940.

Butson, A. R. C., Acclimatization to cold in the Antarctic, *Nature London, 163,* 132–133, 1949.

Cattell, R. B., *Handbook of Multivariate Experimental Psychology,* pp. 1–959, Rand McNally, Chicago, Ill., 1966.

Cherry-Garrard, A., *The Worst Journey in the World,* pp. 1–612, Chatto and Windus, London, 1952.

Davis, T. R. A., Acclimatization to cold in man, in *Temperature–Its Measurement and Control in Science and Industry,* vol. 3, edited by C. M. Herzfeld, part 3, *Biology and Medicine,* edited by J. D. Hardy, pp. 443–452, Reinhold, New York, 1963.

Depocas, F., The calorigenic response of cold-acclimatized white rats to infused noradrenaline, *Can. J. Biochem. Physiol., 38,* 107–114, 1960.

Driver, A. F. M., Physiological characteristics in relation to climatic preference, *J. Appl. Physiol., 13,* 430–434, 1958.

Durnin, J. V. G. A., Exercise, in *The Physiology of Human Survival,* edited by O. G. Edholm and A. L. Bacharach, pp. 301–326, Academic, New York, 1965.

Easty, D. L., Food intake in Antarctica, *Brit. J. Nutr., 21,* 7–15, 1967.

Edholm O. G., Physiological research at British Antarctic Survey stations, in *Human Adaptability to Antarctic Conditions, Antarctic Res. Ser.,* vol. 22, edited by E. K. E. Gunderson, this volume, AGU, Washington, D. C., 1974.

Edholm, O. G., and H. E. Lewis, Terrestrial animals in cold: Man in polar regions, in *Handbook of Physiology,* Sect. 4, *Adaptation to the Environment,* edited by D. B. Dill, E. F. Adolph, and C. G. Wilber, pp. 435–446, American Physiological Society, Washington, D. C., 1964.

Elkington, E. J., Finger blood flow in Antarctica, *J. Physiol. London, 199,* 1–10, 1968.

Elsner, R. W., J. D. Nelms, and L. Irving, Circulation of heat to the hands of arctic Indians, *J. Appl. Physiol., 15,* 662–666, 1960.

Fox, R. H., R. Goldsmith, D. J. Kidd, and H. E. Lewis, Acclimatization to heat in man by controlled elevation of body temperature, *J. Physiol. London, 166,* 530–547, 1963.

Frazier, R. G., Acclimatization and the effects of cold on the human body as observed at Little America 3 on the United States Antarctic Service Expedition 1939-41, *Proc. Amer. Phil. Soc., 89,* 249–255, 1945.

Goldsmith, R., The Commonwealth Trans-Antarctic Expedition–Medical and physiological aspects of the advance party, *Lancet, 1,* 741–744, 1959.

Goldsmith, R., Use of clothing records to demonstrate acclimatization to cold in man, *J. Appl. Physiol., 15,* 776–780, 1960.

Goldsmith, R., and H. E. Lewis, Polar expeditions as human laboratories, *J. Occup. Med., 2,* 118–122, 1960.

Grayson, J., Observations on blood flow in human intestine, *Brit. Med. J., 2,* 1465–1470, 1950.

Grayson, J., Observations on the temperature of the human rectum, *Brit. Med. J., 2,* 1379–1382, 1951.

Hampton, I. F. G., Local acclimatization of the hands to prolonged cold exposure in the Antarctic, *Brit. Antarctic Surv. Bull. 19,* 9–56, 1969.

Heberling, E. J., and T. Adams, Relation of changing levels of physical fitness to human cold acclimatization, *J. Appl. Physiol., 16,* 226–230, 1961.

Héroux, O., F. Depocas, and J. S. Hart, Comparison between seasonal and thermal acclimatization in white rats, 1, Metabolic and insulative changes, *Can. J. Biochem. Physiol., 37,* 473–478, 1959.

Hicks, K. E., Changes in blood-clotting time, serum-cholesterol level, and plasma-prothrombin index in Antarctica, *Lancet, 1,* 30–32, 1965.

Hicks, K. E., Body weight, skin-fold thickness, blood pressure, pulse rate and oral temperature in Antarctica, *Med. J. Aust., 1,* 86–90, 1966.

Hicks, K. E., Changes in the blood-clotting mechanism, serum lipids, and basal blood pressure in Antarctica, *Clin. Sci., 33,* 527–538, 1967.

Hindmarsh, M. E., and R. K. Macpherson, Thermal comfort in Australia, *Aust. J. Sci., 24,* 335–339, 1962.

Hsieh, A. C. L., and L. D. Carlson, Role of adrenaline and noradrenaline in chemical regulation of heat production, *Amer. J. Physiol., 190,* 243–246, 1957.

Joy, R. J. T., J. C. Matone, G. W. Newcomb, and W. C. Bradford, Responses of cold-acclimatized men to infused norepinephrine, *J. Appl. Physiol., 18,* 1209–1212, 1963.

Keatinge, W. R., The effect of repeated daily exposure to cold and of improved physical fitness on the metabolic and vascular response to cold air, *J. Physiol. London, 157,* 209–220, 1961.

Kenshalo, D. R., and J. P. Nafe, The peripheral basis of temperature sensitivity in man, in *Temperature–Its Measurement and Control in Science and Industry,* vol. 3, edited by C. M. Herzfeld, part 3, *Biology and Medicine,* edited by J. D. Hardy, pp. 231–238, Reinhold, New York, 1963.

Khosla, T., and C. R. Lowe, Arterial pressure and arm circumference, *Brit. J. Prev. Soc. Med., 19,* 159–163, 1965.

Kleitman, N., and A. Ramsaroop, Periodicity in body temperature and heart rate, *Endocrinology, 43,* 1–20, 1948.

Law, P. G., Techniques of living, transport and communication, in *Antarctica,* edited by T. Hatherton, pp. 55–80, Methuen, London, 1965.

Law, P. G., and J. M. Béchervaise, *ANARE–Australia's Antarctic Outposts,* pp. 1–152, Oxford University Press, New York, 1957.

Lewis, H. E., J. P. Masterton, and S. Rosenbaum, Body weight and skinfold thickness of men on a polar expedition, *Clin. Sci., 19,* 551–561, 1960.

Lewis, H. E., J. P. Masterton, and S. Rosenbaum, Stability of basal metabolic rate on a polar expedition, *J. Appl. Physiol., 16,* 397–400, 1961.

Lind, A. R., and D. E. Bass, Optimal exposure time for development of acclimatization to heat, *Fed. Proc. Fed. Amer. Soc. Exp. Biol., 22,* 704–708, 1963.

Lowe, C. R., Arterial pressure, physique, and occupation, *Brit. J. Prev. Soc. Med., 18,* 115–124, 1964.

Lugg, D. J., Thermal comfort in Antarctica, *Med. J. Aust., 2,* 746–750, 1965.

Mackworth, N. H., Finger numbness in very cold winds, *J. Appl. Physiol., 5,* 533–543, 1953.

Mackworth, N. H., Cold acclimatization and finger numbness, *Proc. Roy. Soc., Ser. B, 143,* 392–407, 1955.

Macpherson, R. K., The assessment of the thermal environment, A review, *Brit. J. Ind. Med., 19,* 151–164, 1962.

Massey, P. M. O., Acclimatisation to cold in Antarctica, *APU Rep. 262/56,* 40 pp., Appl. Psychol. Res. Unit, Med. Res. Counc., Cambridge, England, 1956.

Massey, P. M. O., Finger numbness and temperature in Antarctica, *J. Appl. Physiol., 14,* 616–620, 1959.

Milan, F. A., Maintenance of thermal balance in arctic Eskimos and antarctic sojourners, in *Antarctic Biology,* edited by R. Carrick, M. Holdgate, and J. Prévost, pp. 529–534, Hermann, Paris, 1964.

Milan, F. A., and K. Rodahl, Caloric requirements of man in the Antarctic, *J. Nutr., 75,* 152–156, 1961.

Milan, F. A., R. W. Elsner, and K. Rodahl, Thermal and metabolic responses of men in the Antarctic to a standard cold stress, *J. Appl. Physiol., 16,* 401–404, 1961.

Nansen, F., *The First Crossing of Greenland,* translated from Norwegian by H. M. Gepp, pp. 1–452, Longmans, Green, New York, 1895.

Nelms, J. D., and D. J. G. Soper, Cold vasodilatation and cold acclimatization in the hands of British fish filleters, *J. Appl. Physiol., 17,* 444–448, 1962.

Nielsen, M., Die regulation der körtemperatur bei muskelarbeit, *Skand. Arch. Physiol., 79,* 193–230, 1938.

Norman, J. N., Cold exposure and patterns of activity at a polar station, *Brit. Antarctic Surv. Bull. 6,* 1–13, 1965.

Orr, N. W. M., Food requirements and weight changes of men on antarctic expeditions, *Brit. J. Nutr., 19,* 79–91, 1965.

Pace, N., Physiological studies in the Antarctic, in *Cold Injury, Transactions of the Sixth Conference, 1958,* edited by S. M. Horvath, pp. 141–174, Macy, New York, 1960.

Palmai, G., Thermal comfort and acclimatization to cold in a subantarctic environment, *Med. J. Aust., 1,* 9–12, 1962*a*.

Palmai, G., Skin-fold thickness in relation to body weight and arterial blood pressure, *Med. J. Aust., 2,* 13–15, 1962*b*.

Palmai, G., Diurnal and seasonal variations in deep body temperature, *Med. J. Aust., 2,* 989–991, 1962*c*.

Pickering, G. W., J. A. F. Roberts, and G. S. C. Sowry, The aetiology of essential hypertension, 3, The effect of correcting for arm circumference on the growth rate of arterial pressure with age, *Clin. Sci., 13,* 267–271, 1954.

Ragan, C., and J. Bordley, The accuracy of clinical measurements of arterial blood pressure, with a note on the auscultatory gap, *Bull. Johns Hopkins Hosp., 69,* 504–528, 1941.

Scholander, P. F., H. T. Hammel, K. Lange Andersen, and Y. Løyning, Metabolic acclimation to cold in man, *J. Appl. Physiol., 12,* 1–8, 1958.

Siple, P. A., and C. F. Passel, Measurements of dry atmospheric cooling in subfreezing temperatures, *Proc. Amer. Phil. Soc., 89,* 177–199, 1945.

Snedecor, G. W., and W. G. Cochran, *Statistical Methods,* pp. 1–593, Iowa State University Press, Ames, 1967.

Soucek, Z., Report on physiology–Wilkes 1962, Antarctic Div., Dep. of Supply, Melbourne, 1963.

Staquet, M., Besoins énergétiques et activité cortico-surrénale chez les membres d'une expédition polaire, *Arch. Belg. Med. Soc. Hyg. Med. Trav. Med. Leg., 19,* 661–712, 1961.

Swan, R. A., *Australia in the Antarctic,* pp. 1–432, Melbourne University Press, Melbourne, 1961.

Teichner, W. H., Assessment of mear body surface temperature, *J. Appl. Physiol., 12,* 169–176, 1958.

van Dilla, M. A., R. Day, and P. A. Siple, Special problem of hands, in *Physiology of Heat Regulation and the Science of Clothing,* edited by L. H. Newburgh, pp. 374–386, Saunders, Philadelphia, 1949.

Wilkins, D. C., Acclimation to heat in the Antarctic, in *Polar Human Biology,* edited by O. G. Edholm and E. K. E. Gunderson, pp. 171–181, Heinemann Medical Books, London, 1973.

Wilson, E. A., *Diary of the 'Discovery' Expedition,* pp. 1–416, Blandford, London, 1966.

Wilson, O., Changes in body-weight of men in the Antarctic, *Brit. J. Nutr., 14,* 391–401, 1960.

Wilson, O., Human adaptation to life in Antarctica, in *Biogeography and Ecology in Antarctica, Monogr. Biol.,* vol. 15, edited by J. van Meighem, P. van Oye, and J. Schell, pp. 690–752, W. Junk, The Hague, Netherlands, 1965.

Winslow, C.-E. A., L. P. Herrington, and A. P. Gagge, Relations between atmospheric conditions, physiological reactions and sensations of pleasantness, *Amer. J. Hyg., 26,* 103–115, 1937.

Wolff, H. S., A knitted wire fabric for measuring mean skin temperature or for body heating, *J. Physiol. London, 142,* 1–2*P*, 1958.

Wyndham, C. H., and H. Loots, Responses to cold during a year in Antarctica, *J. Appl. Physiol., 27,* 696–700, 1969.

Wyndham, C. H., and R. Plotkin, A study of ethnic differences in physiological reactions during acute exposure to cold, and of adaptation of one ethnic group on longer exposure, *Polar Rec., 11,* 500–501, 1963.

Wyndham, C. H., R. Plotkin, and A. Munro, Physiological reactions to cold of men in the Antarctic, *J. Appl. Physiol., 19,* 593–597, 1964.

Yoshimura, H., and T. Iida, Studies on the reactivity of skin vessels to extreme cold, 2, Factors governing the individual differences of the reactivity, or the resistance against frost-bite, *Jap. J. Physiol., 2,* 177–185, 1952.

PHYSIOLOGICAL AND PSYCHOLOGICAL STUDIES CONDUCTED BY CONTINENTAL EUROPEAN AND JAPANESE EXPEDITIONS

J. Rivolier

Service Santé Terres Australes et Antarctiques Françaises and Expéditions Polaires Françaises, Paris, France

Abstract. Biomedical research has been conducted at French stations on the Adélie Coast of Antarctica (primarily Dumont d'Urville) and on Iles Kerguelen, Crozet, and Amsterdam. Studies are grouped under three major headings: psychological, physiological, and biometeorological. Antarctic personnel are examined by interviews and psychological tests prior to departing for Antarctica, and studies of behavioral and medical changes are conducted by medical officers during the winter months in Antarctica. Physiological and medical studies consider peripheral acclimatization, body weight, fatigue, hematology, nutrition, protective clothing, and infectious and traumatic agents. Biometeorological research involves climatological and environmental measurements using sophisticated instrumentation and correlation of changes in atmospheric and environmental conditions with psychological and psychosomatic phenomena. Research at Kerguelen, Crozet, and Amsterdam islands parallels the antarctic studies and offers a variety of environmental and social conditions for detailed study. Physiological studies at Belgian, Swedish, and Japanese antarctic stations also are briefly surveyed.

There are four permanent French stations in the Antarctic and the Subantarctic. Dumont d'Urville is located on the Adélie Coast of Antarctica 66°40′S, 140°01′E. Permanent bases have been established also on Iles Kerguelen, Crozet, and Amsterdam.

The Expéditions Polaires Françaises (EPF) is the organization in charge of the technical aspects of operations conducted on the Adélie Coast under authority delegated by the Ministère de la France d'Outre-Mer, Territoires des Terres Australes et Antarctiques Françaises (TAAF). TAAF is in sole charge of operations at the other three stations.

Besides the projects on the Adélie Coast, EPF is involved with expeditions in other parts of Antarctica and in the Arctic, notably with the Soviet Antarctic Expedition, the projected International Glaciological Expedition in Antarctica, and the International Glaciological Expedition to Greenland.

The national organization representing the Scientific Committee on Antarctic Research (SCAR) in France is the Comité National Français des Recherches Antarctiques (CNFRA). A representative for medical and human biological research is on this committee. CNFRA evaluates research programs and submits them to the Scientific Council of TAAF, which also has a medical representative. The Scientific Council and the director of the scientific laboratories of TAAF authorize, supervise, and assist the programs, depending on their interest and available logistic and financial support. Medical research at the four French stations and in other expeditions is presently the responsibility of the author. Preemployment medical examinations of participants are carried out with the assistance of the specialized services of the Service de Santé des Armées (Army Medical Service).

FRENCH STATIONS ON THE ADELIE COAST

French exploration and research began with the voyage of the MS *Commandant Charcot* in 1948–1949. The first French wintering-over party on the antarctic continent was located at Port Martin (64°49′S, 141°24′E) during 1949–1951 and had a complement of 11 men. During the next year (1950–1952), 17 men wintered at Port Martin, which was destroyed by fire in 1952.

Small parties wintered on Pétrel Island (Ile des Pétrels) (66°40′S, 140°01′E) during 1951–1953 (7 men) and 1955–1957 (14 men). A new base, Dumont d'Urville, was built on Pétrel Island, adjacent to the mainland coast, during 1957 in preparation for the International Geophysical Year.

Two small parties of 3 men wintered over at Charcot station, located on the South Polar Plateau, 2400 meters above sea level (69°22′S, 139°01′E), during 1956–1959.

A new station on the South Polar Plateau is planned for the future. It will probably be situated about 500 km inland from the Adélie Coast.

The years of operation and personnel complements for Dumont d'Urville station were as follows:

Year	Winter Party
1956–1958	20
1957–1959	20
1958–1960	12
1959–1961	14
1960–1962	18
1961–1963	19
1962–1964	20
1963–1965	19
1964–1966	20
1965–1967	29
1966–1968	27
1967–1969	27
1968–1970	27

The climate of the Adélie Coast area where Dumont d'Urville is located, although it is essentially polar with summer temperatures below freezing, is tempered by its proximity to the sea. Temperatures do not fall below -40°C. Winds are very strong and frequent; velocities of more than 200 km/hr have been recorded.

Supply and relief of wintering parties are carried out entirely by sea during the austral summer between mid-December and mid-February.

The station is composed of many service buildings and laboratories; the process of equipping the station is still in progress. At present the medical section has only one room for use as a laboratory, but medical equipment is located also in several other places. Room for a sick bay and a laboratory is provided in new residential buildings under construction.

Medical research is carried out by the wintering medical officer, who can call on an electronic engineer and meteorologist for technical assistance. In 1960–1962, two medical officers, one of them a psychiatrist, wintered over at Dumont d'Urville. Future program plans provide for biologists to participate with the medical officers in carrying out medical research.

Medical research on the Adélie Coast began on the initiative of medical officers wintering there before there was any definite policy on research. Certain early research studies are mentioned because they sometimes provided useful initial data for planning the more systematic research programs developed later by the responsible organization in Paris.

Research studies are grouped under the following headings: psychological, biophysiological, and biometeorological.

Psychological Research

From the first winter expeditions it became evident that the psychological structure of the group, as well as the behavior of the individual, underwent modification during an antarctic sojourn. The expedition medical officers, trained primarily in surgery, felt ill equipped to study these phenomena but recorded the principal clinical manifestations and tried to determine their causes [*Rivolier,* 1954; *Sapin-Jaloustre,* 1955; *Goy,* 1959]. These first publications mentioned possible correlations between psychological tensions and meteorological factors.

Aware of the practical psychological adjustment problems of wintering over [*Rivolier,* 1954], EPF recommended that research be carried out in this field, especially in relation to personnel selection. The neuropsychiatric service for airborne personnel of the French Air Force agreed to examine candidates for the expeditions by means of interviews and a series of tests. For several years, candidates were subjected to these examinations, but the examinations were not used in selection in order to determine whether the prognosis made before the subjects' departure was in fact borne out by the notes made by the medical officer during the stay in Antarctica. During the winter months the medical officer also administered certain projective tests and carried out sociopsychological investigations. The subjects were again interviewed and tested on their return to France [*Soussen and Digo,* 1964]. After the specialists conducting this work had gained experience over several years, it was decided to take into account the results of the initial psychological examinations at the time of hiring.

During the 1960–1962 expedition it was possible for a psychiatrist to spend the winter and carry out psychological, psychosomatic, sociological, and psychopathological investigations [*Perrier,* 1964, 1965, 1967; *Rivolier and Perrier,* 1966].

In addition to making direct observations, it was possible to study psychological changes in the wintering staff by indirect methods. These included the analysis of personal consultations (motivational problems, recurrent adjustment difficulties, etc.), psychosomatic symptoms (asthenias, headaches, sleep disorders, digestive syndromes, etc.), and manifestations of individual or group irritability (character difficulties). In addition to noting the psychological behavior of individuals, an attempt was made to study group behavior and the relationships between the group and the local authority (the station leader) and the authority of the supporting organization in Paris. Many other aspects of aggressive behavior, direct or indirect, immediate or secondary, resulting in individual or group disturbances also were studied.

From direct observations during the wintering period, new data were obtained on homosexuality in a polar community. This work dealt primarily with homosexual tendencies or equivalents and with possibilities for personnel selection based on studies of the masculinity-femininity factor of the Minnesota multiphasic personality test [*Perrier*, 1965].

During the winter of 1961 the first correlational studies were conducted [*Rivolier and Perrier*, 1966]. Since these initial studies were, admittedly, on a very limited scale, it was considered desirable to follow up and improve on them in future expeditions. These investigations involved the correlation of meteorological parameters with character disturbances and physical complaints. Standard daily meteorological data, grouped into periods depending on the nature of the weather and a daily weather index, were compared with daily records of character disturbances and physical complaints recorded objectively during a clinical consultation. The agreement between the incidence of character disturbances and the incidence of both physical complaints and various states of the weather or its components was noted. The conclusions are summarized below.

1. No correlation with outside temperature was present apart from an unfavorable psychological effect from temporary recurrences of low temperatures during spring and summer.

2. The incidence of physical complaints and character disturbances was correlated with wind velocity, especially when the wind was strong; the incidence of character disturbances was highest during a blizzard.

3. Both physical complaints and character disturbances tended to vary with atmospheric pressure, but the relationship was probably indirect.

4. There was an ill-defined relationship with solar radiation.

5. There was a lack of synchronization between the intensity of disturbances and the maximum intensity of meteorological phenomena.

6. Concerning the type of weather, it was found that the most characteristic days (good or bad weather) were those with the most disturbances and complaints; although there was a perceptible increase in the number of physical complaints during bad weather, a definite increase in character disturbances was noted during both good weather and bad weather.

To sum up, it can be said that there are certain correlations between psychomedical manifestations and the biometeorological environment (e.g., wind velocity, changes in weather, and periods of characteristic weather and probably others). The psychological environment also probably plays a major role, and clinical phenomena are certainly related to it, for they are the commonest expressions of psychosomatic relationships; in this study, however, such a relationship was not indicated by a correlation between the number of character disturbances and the number of daily consultations. The biometeorological environment possibly plays an indirect role by modifying the quality of the psychological environment and, in turn, produces abnormal psychosomatic and behavioral reactions.

The Conference on Medicine and Public Health in Arctic and Antarctic Regions, held in Geneva in 1962 under the auspices of the World Health Organization, noted the importance of psychological problems during the winter period, which constitute the major medical problems at antarctic stations, and the need to intensify their study [*Rivolier*, 1962*a*].

In the organization of French expeditions an integrated scheme was put into effect to cover both selection and psychological research. A type of feedback was established in which research helped to establish the best methods of selection, and selection constituted an integral part of the research. This important scheme has never been described in French literature and has been made known only in specialized media, such as the SCAR Subcommittee on Human Adaptability Programs in the Antarctic. It will be useful to refer to it extensively here.

Before continuing with this point, it is stressed that our experience has shown that all psychological examinations requiring the collaboration of the subjects are possible only before the departure. During the winter the subjects cannot be asked to participate in psychometric tests other than proficiency tests. Personality tests and sociometric methods are accepted very reluctantly or not at all, and results based on the responses of only those who will cooperate are valueless. It is also impossible to contact the members after they have returned: most individuals lose contact with the employing organization, shun any kind of supervision, and resume family life. It is normal and not surprising that these men should disappear from the scene.

Selection. Predeparture examinations, carried out officially with the aim of selection, yield useful scientific data for psychological research provided that standardization permitting eventual statistical analyses is adopted. These examinations have two objectives; one is negative in that those with low intelligence, psychoneurotic traits, temperamental unsuitability, alcoholism, marked cyclothymic traits, manifest homosexual tendencies, etc., can be eliminated as being unsuitable, and the other is positive in that one or another type of personality can be chosen on the basis of preferential criteria.

The evaluations are carried out by three examiners; two use interviews, and the third uses a series of tests. The interview is carried out in two parts, the first by an examiner who is a specialist in psychiatry and psychoanalysis and the second by a physician who is not a specialist in psychiatry but who has had considerable polar experience. The interview, which can be very flexible, must attempt to bring to light information concerning occupational adjustment, the quality of emotional responses, defensive reactions and temperament (introversion or extroversion), motivation, family problems, the degree of maturity, depressive and hypochondriacal tendencies, paranoid ideation, the attitude toward authority, the ability to adapt socially, and so on. At the end of the examination the examiner prepares a report covering five principal areas: (1) attitude (the subject's attitude toward the selection situation and his reaction to the examiner's neutral attitude), (2) intelligence (capacities for judgment, reasoning, and expression rather than an intelligence quotient), (3) biography (logical sequence in reporting biographical events related to occupation and family), (4) motivation (attempt to overcome the candidate's reticence and infer deeper levels of motivation), and (5) personality (evaluation of personality traits manifested in the candidate's attitude, conversation, biography, and his emotional responses).

An attempt is then made to assess the quality of the candidate's psychological defense mechanisms and his adaptability under conditions of frustration and defensive reactions brought about by wintering over in the Antarctic.

After a number of trials revealing difficulties in the use of particular tests the following set of tests was adopted. For IQ tests the Benoit-Pichot vocabulary test and test D.48 are used. The candidate is instructed to write two portraits, i.e., two descriptions of himself, one by a 'good friend' and one by a harsh critic. For projective tests the Van Lennep picture story method, in which the subject writes stories describing relatively stereotyped situations portrayed in four photographs, is used. The Rorschach test is also administered. A personality questionnaire, the Minnesota multiphasic personality inventory, is administered.

After the selection examinations are completed, the candidates are classified into four groups in terms of acceptability: group A, acceptable; group B, acceptable but with restrictions on particular points; group C, unacceptable unless a replacement cannot be found; group D, unacceptable under any circumstances.

Research. The research data comprise information collected during the selection procedure, during the winter, and after the winter has ended.

The aim of psychological research during the winter is to obtain an objective assessment of the adaptation of every member of the expedition. Because no testing of subjects is carried out during their stay, studies of observed behavior rather than analyses of questionnaire responses are conducted. The medical officer responsible for psychological research must act as a neutral observer, and he is required to record individual and group reactions to the situation.

The areas of study are considered below. The factors affecting the behavior of station members have been classified schematically under internal factors, i.e., those generated by the station environment during the winter, and external factors, i.e., those arising from events in the outside world. Internal factors can be subdivided into individual and group factors. Individual factors include: (1) social factors (aggressive behavior and conflicts between peers), (2) subordination (relationships with direct superiors and the station leader), (3) work factors (difficulties with equipment or apparatus or disappointment concerning position or work), (4) character factors (psychasthenic fluctuations attributable to the individual's personality), (5) biometeorological factors (grossly abnormal sensitivity to the weather), and (6) catastrophic events (personal feelings about notable crises at the station, such as fire or a helicopter crash). Group factors include: (1) social factors (intergroup relationships), (2) subordination (relationships of the group with the immediate superior or station leader and group reactions), (3) group work factors (problems arising during group work, such as fatigue and routine duties), (4) biometeorological factors (meteorological events with repercussions affecting the group), and (5) catastrophic events (notable events felt by the group as a whole).

External factors also can be subdivided into individual and group factors. Individual factors include: (1) emotions and character (essentially personal relationships with the family), (2) subordination (relationships with EPF organization and research directors), and (3) external world events (political events, wars, etc.). Group factors are: (1) subordination (relationships between the group or expedition as a whole and EPF organization and research directors) and (2) external events (those world events with an impact on the group or the expedition as a whole).

These factors appear in the daily record of character disturbances that is filled out by the medical officer during the winter.

Possibilities for study during the winter include the observer's daily notes concerning psychological disturbances of individuals and groups together with possible causes. The names of persons or groups involved in these disturbances are specified. Gross changes in the behavior of each individual can be indicated by asking him to assess the quality of the day's experience by means of a questionnaire. Every evening the individual is asked if he

BIMONTHLY PSYCHOLOGICAL CHECKUP

NAME: ______________________ DATE: ______________

	Response Categories				
MOOD	Depressed	Unhappy	Normal	Euphoric	Too gay
SLEEP	Poor	Little	Normal	Much	Excessive
APPETITE	Poor	Little	Normal	Large	Excessive
ALCOHOL	None	Little	Normal	Much	Excessive
CONVERSATION	None	Little	Normal	Much	Excessive
GAMES	None	Little	Normal	Much	Excessive
READING	None	Little	Normal	Much	Excessive
WALKING	None	Little	Normal	Much	Excessive
WORK IN HIS OWN FIELD	None or poor	Little	Normal	Much	Excessive
GROUP WORK	Absent or awkward	Passive	Adapted	Active	Leader
GROUP LEISURE ACTIVITIES	Absent or against	Passive	Adapted	Good team member	Organizer
DOMESTIC SERVICE	Poor	Relaxed	Normal	Good	Excellent
COMPANION	Poor	Difficult	Average	Good	Excellent
SOCIABLE	No	Little	Average	Yes	Very
ATTITUDE TOWARD LEADER	Aggressive	Passive	Normal	Supporter	Fawning
ROLE IN GROUP	Scapegoat	Retiring	Adapted	Active	Leader
ATTITUDE TOWARD GROUP	Aggressive	Passive	Cooperative	Active	Leader
ATTITUDE TOWARD MEDICAL OFFICER	Aggressive	Passive	Ordinary	Cooperative	Fawning

CHECK THE APPROPRIATE CATEGORY.

Fig. 1. Bimonthly psychological checkup.

has had a good day, and he responds yes or no. The replies provide global and subjective impressions. Every two weeks, various aspects of each subject's behavior can be assessed by filling out the bimonthly psychological checkup (Figure 1). The normal response is the norm for the subject examined, that is, the individual's average state. This form includes items reflecting mental status on the one hand and relations with the group on the other.

Some behavioral disturbances reflect difficulties of adaptation (hypersomnia or insomnia, anorexia or bulimia, fatigue unrelated to work effort, and withdrawal or absenteeism from group activities such as meals and amusements) and will appear both in the bimonthly psychological checkup and in consultation records.

A general history reflecting the psychological evaluation of the group can be compiled by means of a concrete and objective 'diary.' Both the individual and the collective development can be specified without detailed indications. The diary, which is a record of events, their psychological effects, the data collected, and the development of individual relations and which is also an indicator of the observer's own reactions and mental state, can provide only a framework within which it will be possible to fill in finer details and explanations at a later time.

At systematic monthly examinations and at personal consultations sought by members of the winter expedition the physician-observer has the opportunity to gain added knowledge of the individuals and of the group. Repeated consultations in particular demand an interpretation of the motives for seeking the consultation and of their consequences, in other words, the psychological basis for the complaint motivating the seeking of medical advice. An example of the layout of a consultation record is shown in Figure 2. In particular, it should clearly be indicated whether the consultation is concerned with an organic or a functional condition.

Individual or group phenomena can sometimes be precipitated or influenced by meteorological factors. It is important to note the meteorological conditions daily on two different levels, the subjective level, describing 'what sort of weather it seems to be,' and the objective level, consisting of official meteorological data indicating the local climatic conditions. This does not include special biometeorological information supplied by instruments under the medical officer's charge. The form used to record this information is shown in Figure 3. Note that this meteorological information is used for purposes other than psychological correlation.

After the participants have returned from an expedition, it is impossible to contact the subjects for further examination. The medical officer and the expedition leader are asked to complete a wintering-over evaluation (Figure 4) 3 months after returning to France, this evaluation thus serving as a retrospective assessment of the

CONSULTATION RECORD

PART 1 DAY: MONTH:

Name and job of person seeking advice:	Date and Times:		Motive:	Symptoms:	Signs:
	Complaints:	Consultation			

PART 2 YEAR:

Treatment given:	Mood	Sleep	Stopped work:	Date of resumption of effective work:	Objective or functional:	1st consultation or follow-up:

Fig. 2. Consultation record.

characteristics of each member of the expedition during the winter months and providing a check on the selection procedures.

Biophysiological Research

This research deals with pathology and with selection, protection, and adaptation problems. During the early expeditions the medical officers were given the task of observing their companions closely, drawing up research forms, and deducing practical conclusions. Certain preliminary investigations without any concerted direction also were carried out. Since that time, methodical studies have been undertaken, and this work has resulted in several publications. Most importantly, this program has permitted systematic collection of data over a number of years for statistical analysis. This work is being carried out in the spirit of the SCAR Subcommittee on Human Adaptability Programs in the Antarctic. The recommendations of the SCAR subcommittee involve the application of simple techniques that can be carried out by expedition medical officers and have the advantage of permitting repetition and providing a common program for various countries participating in SCAR.

Physiological adaptation. Two series of experiments [*Rivolier*, 1953; *Sapin-Jaloustre*, 1955] on cooling and cutaneous adaptation were conducted on the Adélie Coast. Metal heat loss calorimeters with a capacity from 0.5 to 1 liter of water were used. One of the instruments was tested in the laboratory in Paris, and as a result the theoretical cooling curves were plotted, and a formula was deduced for the relationship between wind velocity and cooling time at specific ambient temperatures. This numerical expression was considered the convection coefficient of the instrument [*Sapin-Jaloustre*, 1955].

In the Antarctic the cooling times of continuously agitated liquid between given temperatures, for example, between 37°C and 0°C or between 40°C and 20°C, were studied. Measurements made under varied weather conditions, e.g., at low temperatures, without wind and in the presence of wind, and during blizzards, indicated the great importance of wind in heat loss, especially during blizzard conditions.

Several attempts were made to estimate cutaneous, that is, peripheral, adaptation [*Rivolier*, 1953]. The following experimental procedure was used for this purpose. With the subject recumbent and at rest the skin temperature of both hands, the pulse rate, the arterial pressure and its oscillations, and the skin color were noted. One hand was cooled in a bath at the temperature of melting ice for 1 min. The same measurements were repeated at the end of immersion. The shape of the skin temperature curves and the amplitude and duration of the oscillometric curves were studied. The time taken for temperature values to return to normal also was noted. Examination of these curves has shown that in some wintering subjects the cutaneous vasomotor reactions of the hands improved after a few months. This phenomenon can be regarded as peripheral acclimatization.

The body weight of wintering subjects has been investigated systematically during most expeditions. In general, these studies have revealed different mean values during three periods. During the summer, soon after arrival at the base, body weight generally decreases. During the following winter months, roughly extending from March to September or October, weight tends to increase. During the third period, corresponding to the end of winter or spring, body weight decreases.

These commonplace observations of body weight are readily explained by the pattern of life and work of

BIOMETEOROLOGICAL DATA

Base: DUMONT d'URVILLE YEAR: MONTH:

Date	Temperature, °C			Atm. Pressure			Rel. Humidity			Ave. Wind		Sunshine	Notable meteor-ological phenomenon	Remarks
	Max.	Min.	Mean	Max. mm	Min. mm	Mean mm	Max.	Min.	Mean	Vel.	Dir.	h + h/10		

Fig. 3. Biometeorological data record.

wintering subjects. During the winter they have relatively little physical activity, consisting mostly of indoor work, and they tend to overeat grossly. During the summer and spring, on the other hand, station members are much more active, this activity resulting in a diminishing weight curve, or a return to normal body weight.

Measurements of skin folds failed to reveal any parallel with body weight. This result might have been partly due to variation or unreliability in the measurement of skin folds among different experimenters.

In a series of winter expeditions, fatigue was studied in all members by a combination of clinical observations and tests. In particular, an effort tolerance test (Schneider's test) was used, which is especially adapted to this type of study because it reveals signs of fatigue in nonathletic subjects, or more precisely in subjects without particular physical training; the sensitivity of the test is very high for such subjects.

The results of this study [*Rivolier et al.*, 1967] showed that in more than 50% of the subjects the Schneider test clearly gave decreased values of effort tolerance in June and July (midwinter); in another 30% of the subjects, values on the Schneider test decreased by October–November. It is curious to note that the marked decrease in values on the Schneider test around midwinter did not correspond to decreases in body weights, whereas in the autumn, for the most part, values on the Schneider test decrease at the same time as body weights. Also, there was no correlation between lower values on the Schneider test at midwinter and manifestations noticed by the subjects; that is, the clinical manifestations of fatigue reported subjectively were aberrant in relation to the test values.

An explanation of the lowering of values on the Schneider test in winter might be sought in terms of the pattern of life at the station. Work cannot be responsible because it is lightest at this time of year. Two factors are notable during the winter period. First, the disappearance of daylight and the absence of solar radiation might relate to fatigue unaccompanied by physical deterioration. Second, it is during the midwinter period that the morale of expedition members is most difficult to sustain. With the return of the sun in the following months, despite individual difficulties, morale is better because the subjects are approaching the end of their ordeal. It could be hypothesized therefore that, during midwinter, physical apathy and psychological asthenia, related to psychosomatic causes, may account for the low values on the Schneider test. The lowering of values on the Schneider test in the autumn can adequately be explained by a resurgence of activity and work.

Measurements of neuromuscular excitability made with the aid of an electronic rheotome did not yield any significant results except in the course of a traverse expedition [*Nègre*, 1960].

In hematologic measurements carried out on winter expedition members [*Artignan and Guériot*, 1971], marked increases in red cell count were found during the first half of the winter period, the count reaching $6 \times 10^6/mm^3$ at about midwinter in some subjects. A progressive decrease in the red cell count during the second half of the winter period resulted in a return to the initial values.

The white cell count showed a rapid initial decrease, followed by a slow, regular decline until the end of winter. In the few cases for which the study continued after the end of the winter, a rapid rise in the white cell count was observed.

The differential white cell count revealed a crossing of the curves for the polymorphs, whose number decreased, and the lymphocytes, whose number increased. The levels of eosinophils and monocytes appeared stable except for an increased percentage in April. An earlier study [*Sapin-Jaloustre*, 1955] carried out on the Adélie Coast failed to reveal an increase in the eosinophil count, this study contradicting other findings.

The erythrocyte sedimentation rate, along with the white cell count, tended to fall immediately after arrival at the base. It remained relatively stable during the winter, rose slowly, and fell again at the end of the winter.

Some of these variations had already been reported by other investigators. Polar anemia, however, frequently described by the medical officers of the first antarctic expeditions, has not reappeared. This condition perhaps

REVIEW OF WINTER EXPEDITION

NAME: ______________________________

Official position: ______________________________

Date of review: ______________________________

1. How did he stand up to the winter sojourn?

Very well	Well	Average	Badly	Very badly

Your assessment:

2. What sort of companion was he?

Very good	Good	Ordinary	Poor	Very poor

Your assessment:

3. How did he carry out his own work?

Very well	Well	Average	Badly	Very badly

Your assessment:

4. How did he join in with the team work?

Very well	Well	Average	Little	Never

Your assessment:

5. Was he a complainer?

Very often	Often	Sometimes	Rarely	Never

Your assessment:

6. a. List in order the persons about whom he complained chiefly:

   b. List in order the things about which he complained and the nature of his complaints:

7. What was his role in the group?

   a. Choose from the following descriptive terms the one or ones corresponding to his role:

leader	solid citizen	serious type	life and soul of the party	
scapegoat	black sheep	active	passive	revolutionary
anarchist	a burden on the group			

   b. Add whatever qualifications are necessary to complete this description:

8. Was he ever unhappy:

Very	Yes	Sometimes	Rarely	Never

Your assessment:

9. Was he ever unable to sleep?

Very often	Yes	Sometimes	Rarely	Never

Your assessment:

10. Was he ever a problem because of excessive drinking of alcohol?

Very often	Yes	Sometimes	Rarely	Never

Your assessment:

11. Was he ever ill (without reference to the drug register)?

Very often	Yes	Sometimes	Rarely	Never

Your assessment:

12. Did he ever say he was tired?

Very often	Yes	Sometimes	Rarely	Never

Your assessment:

13. a. Would you winter with him again at Dumont d'Urville?

Certainly	Yes	If necessary	No	Certainly not

b. If so, in what capacity would you take him?

c. Any comments?

14. Conclusion: All in all, what sort of person would you say he was?

Very good	Good	Average	Bad	Very bad

Have you anything you wish to add?

Date: ____________________

Signature: ____________________

Fig. 4. Review of winter expedition.

was simply a reflection of vitamin or other nutritional deficiencies. The observed polycythemia apparently reflected hemoconcentration, as the elevated hemoglobin estimations and hematocrit values obtained during the winter would suggest. The role of cold in this hemoconcentration remains obscure, but the changes in the differential white cell count can perhaps be explained by the absence of pathogenic microorganisms in the air and the food. It may be that the elevation of the eosinophil count observed in the early expeditions was related to physical or mental stress, giving rise to an intense adrenal response. This situation does not exist now because of the orderly life, generally without stresses, in modern antarctic expeditions.

During one expedition a medical officer measured the growth of nails and hair [*Sapin-Jaloustre*, 1955]. The technique used to measure nail growth was to establish a comparison with the bony skeleton by means of a metal wire placed on the nail and radiographed. The technique used to measure hair growth was to measure the length of hair growing between two shaves during a given interval of time. The results demonstrated a slowing of growth of the skin appendages. It was suggested that this slowdown might vary with the conditions of group life and, in particular, with the atmosphere in which the subjects lived and the temperatures that they had to endure.

Nutrition and clothing protection. Many studies of nutrition have been conducted during French polar expeditions [*Rivolier*, 1953, 1963*b*, 1955*b*, 1967*b*; *Rivolier and Le Bideau*, 1960; *Sapin-Jaloustre*, 1955]. These studies made it possible to determine the quantity of food needed as well as the acceptability of the foods provided. They have obtained practical data for use in planning rations.

Two types of rations can be distinguished, those for traverse parties and those for wintering at the base. The study of the food problems of traverse parties was one of the first concerns of our expeditions. The original expeditions, carried out on foot and with dogsleds, called for light, concentrated, and easily prepared rations. These early studies were nothing more than a fresh look at the problems facing the first expeditions before World War I. The calorific value of the needed ration was accepted to be 4000–4500 calories at the maximum, in contrast to the calorific value of the vast quantities of food provided in the old expeditions. Although the fat content remained high, it was lower than that previously adopted. A practical ration containing about 110–120 grams protein, 120–150 grams fat, and 600–800 grams carbohydrate was established.

Concentrated foods, especially sweetened semifluid foods with high fat content, were tested at various times. In recent years, lyophilized foods have assumed an important role in providing rations for traverse parties. Such expeditions are mechanized today, and the weight restrictions are not so severe as they used to be. Rations issued by EPF [*Rivolier*, 1963*b*] are equivalent to about 4000–4200 calories per man daily and a weight of about 1.5 kg per man daily. In general, rations are made up for three men for 2 days, and 14 different menus have been provided for variety. The canned food taken on expeditions is chosen with great care in regard to quality and preservation properties. Vacuum-packed plastic containers are used for some products, notably for lyophilized items.

Cold tests and laboratory preservation tests have been carried out during several expeditions. Artificial aging procedures in the laboratory have been supplemented by more lifelike studies of foodstuffs kept for several years in polar regions and brought back to France for analysis.

After much controversy it has been shown that the previous scale of rations for wintering at a base was for the most part too generous. Analysis of the amount of food offered and the amount of waste showed that a ration of between 3000 and 3200 calories per man daily was quite adequate for wintering at the base; the ration provided 110–120 grams protein, 100 grams fat, and 450–500 grams carbohydrate.

At the same time, great effort has been made to improve the variety and quality of canned foods offered. Increased use has been made of frozen foods, especially meat, and also fresh foods by providing for climatically controlled food stores.

The need for wines and alcohol on French expeditions was very low over the years. Recently, with improvements in comfort and possibilities for living in different places, alcohol consumption has increased. Living conditions on polar expeditions change constantly, the result being greater demands and needs of individual members and, frequently, increased psychological difficulties as well.

It is the present practice in French expeditions to provide survival kits with rations of 1000 calories, consisting, for example, of pemmican, a semiliquid food, and sugar. In every kit it is obviously necessary to provide some form of fuel so that water can be obtained from ice.

Research conducted since the beginning of French polar expeditions has resulted in the development of effective clothing and equipment for traverse parties along with improvements in living quarters. For life at the base, current conditions are such that conventional equipment can be used [*Rivolier*, 1953; *Sapin-Jaloustre*, 1955]. A diving suit and antiblizzard helmet were tested but did not prove sufficiently useful to warrant general introduction.

In the last few years, extensive laboratory studies of modern materials, fibers, lightweight plastics, polyester insulation layers, and so on have been made. Along with measurements of mechanical resistance and studies of permeability to wind and water vapor, measurements of thermal insulation were made by two methods: (1) determining the insulating capacity by covering of heat loss calorimeter with the test material and (2) detecting transmission from a particular source by means of thermopiles placed on each side of the test material. As a result of these laboratory measurements, recommendations for new equipment were made, and these have been put into effect.

When French mobile expeditions are in the field, they no longer use tents but tractor-drawn caravans, in which members can be relatively comfortable with the aid of forced air heaters.

Improved medical-surgical equipment, knowledge of polar living conditions, and the training of members have reduced the incidence of accidents in French polar expeditions to the minimum and made it possible to formulate effective safety policies, both technical and medical [*Rivolier*, 1957].

The appearance and spread of new diseases at antarctic bases has been described by *Rivolier* [1962*b*].

Medical selection. Before departing for the polar regions, all subjects are given thorough physical examinations at a military hospital [*Rivolier*, 1963*a*, 1964]. The examinations are of the standard type, placing particular emphasis on the cardiovascular system with electrocardiograms (at rest and during exertion), chest X rays, and complementary laboratory tests (blood count and formula, blood sugar, sedimentation rate, bleeding and coagulation times, Bordet-Wassermann reaction, and blood group). In addition, all subjects receive otorhinolaryngological, ophthalomological, and stomatological examinations. These physical examinations reveal contraindications to wintering in the Antarctic and also permit differential selection. Recall that all subjects are medically examined once a month during the sojourn after the biophysiological observations, and these examinations permit comparisons with the results of the physical examination made before the subjects leave on the expedition.

An experimental study [*Rivolier*, 1953] conducted with the previously described cold water test revealed individual differences in peripheral sensitivity to cold. Measurement of skin temperatures and of oscillations after immersion of the hand in cold water and examination of the resulting curves make it possible to eliminate candidates with poor vagosympathetic vasomotor reactions or at least suggest that their adaptation to cold may be unsatisfactory. This research is currently being resumed with up-to-date apparatus.

Pathology. Cases of frostbite encountered in French polar expeditions have always been benign [*Sapin-Jaloustre,* 1956], and none have required specialized treatment.

Infectious diseases are rare in polar expeditions, yet during one 2-year period there was an epidemic of Queensland fever on the Adélie Coast. Because of the dates of the epidemic (autumn), the vector could not be determined: meat imported from Australia or its packing materials seemed to be the most likely etiological factor.

The only conditions requiring surgical treatment during the French expeditions have been commonplace injuries or conditions, including a few cases of acute appendicitis, one case of intestinal obstruction, and two eye injuries.

During one winter expedition, many members were poisoned by soup made of concentrated fresh euphausids, which probably released a type of histamine shock.

Several cases of carbon monoxide poisoning, usually occurring in garages or vehicles, were treated successfully.

No case of canned food poisoning has occurred; note that modern canned foods can ensure freedom from all deficiency diseases.

Biometeorological Research

Early research. During one of the first expeditions to the Adélie Coast an attempt was made with the facilities then available to obtain climatological and biographical data from the Géologie Archipelago [*Rivolier*, 1955*a*]. This first investigation served as the starting point for later extensive work. Studies carried out in France [*Rivolier*, 1967*a*] made it possible to define the biometeorological parameters to be studied and to establish the principal types of apparatus to be used. Psychosomatic correlations were studied during one winter expedition by using only standard meteorological parameters and a weather index [*Rivolier and Perrier*, 1966]. The results of this study are described in this section under psychological research.

Research in progress. The Davos frigorimeter appeared to be the instrument best suited to the study of the cooling power of the environment. One such instrument, modified for polar regions, was installed on the Adélie Coast. Coupled with its continuous, integrated recording system is a device giving the temperature, measured by a resistance thermometer, and the wind velocity, measured by an integrating anemometer. The results of the analysis are then compared with solar radiation measurements made by a solarigraph. Note that, although the amount of electricity supplied to this polar frigorimeter is twice that used for similar instruments in Switzerland, the cooling due to the wind and blizzards is so great that the quantity of current available is insufficient under extreme conditions, and the temperature of the sphere cannot be maintained at a constant level. Modifications were introduced to obtain an instrument resistant to the most extreme conditions. As a result of continuous measurements taken in all kinds of weather, exact relationships between the various heat loss parameters have been established. Following the work of *Siple and Passel* [1945] it will be possible to derive formulas for cooling power based entirely on experimental results and to evaluate the relative importance of each parameter.

Global radiation and diffuse radiation have been studied on the Adélie Coast by means of solarigraphs equipped with Kiep-Zonen thermopiles. An instrument for measuring ultraviolet radiation, an Eppley UV photometer, also has been installed. This instrument seems to be well suited to and sufficiently accurate for general climatological measurements. Radiation measurements will be supplemented by installing solarigraphs measuring reflected radiation and infrared radiation.

An instrument for measuring the electrical field gradient has been installed on the Adélie Coast. This apparatus, recommended by L. Koenigsfeld, consists essentially of an antenna carrying a radioactive element connected to an electronic amplifier with a recorder.

An instrument for measuring atmospheric ionization, providing information on the quality and quantity of negative and positive ions and of small and large ions, will soon be installed.

By means of a resistance thermometer and hygrometric probes, the climatological conditions in the interior spaces of the station at different levels, i.e., near the floor, at the height of a man, and at the ceiling, are being established. The results will provide a better understanding of living conditions at the base and will allow technical improvements to be made. The measurements of ionization described above also will be carried out in the interior of the buildings.

FRENCH STATIONS ON SUBANTARCTIC ISLANDS

The Pointe Molloy station (49°21′S, 70°04′E) in the Iles Kerguelen was opened in 1959 and operated for 1 year only. Port-aux-Français station (49°21′S, 70°12′E) also was opened in 1959 and has operated continuously with a new wintering party every year. Between 90 and 100 persons are present at Port-aux-Français station. The station on Ile de la Possession, located in the Iles Crozet (46°25′S, 51°52′E), has been open since 1965, and wintering parties have been there without interruption until the present. The number of persons wintering at the station is between 15 and 20. La Roche Godon is located on Ile Amsterdam (37°48′S, 77°32′E). The station was opened in 1959 and has operated continuously with 35-40 persons wintering each year.

Iles Kerguelen and Crozet have a climate of the subantarctic type. The temperature rarely falls below −8°C in the winter, and in the summer the temperature is above 0°C but is in the low positive range. The winds are relatively strong, and atmospheric disturbances are very frequent. On Iles Crozet, in particular, there is little sunshine because of frequent fog and abundant precipitation. The snowfall is slight, however, at these stations. The microclimates on different parts of Ile Kerguelen vary considerably depending on geographic location. Large glaciers, descending to the sea, are found on the island. Port-aux-Français station enjoys excellent microclimatic conditions because it is more sheltered from the wind and receives more intense sunshine than the other subantarctic stations.

Ile Amsterdam, which is entirely volcanic, lies on the border line between tropical waters and subpolar waters. Its climate is very mild, having a small temperature range, that is, between 5° and 10°C in winter and reaching 20°C in summer. Winds are fairly strong, and atmospheric disturbances are frequent. In particular, Ile Amsterdam is a meeting place for depressions coming from the south and the tail ends of cyclones coming from the north. Isolated halfway between South Africa and Australia, this island has a typical oceanic climate.

All three islands are supplied once a year in the summer, and the station personnel on the islands spend the winter in much the same way as those at polar bases. The buildings and living conditions on Iles Crozet and Amsterdam can be compared with those at a polar station, such as a station on the Adélie Coast; Port-aux-Français station on Ile Kerguelen is very different, however. It stretches over several kilometers and consists of many residential buildings and laboratories.

Organized medical research is only beginning on the subantarctic islands, and only a few isolated studies have been carried out by wintering medical officers on their own initiative.

Past research. From meteorological data collected over several years on the three islands a study was carried out to define the bioclimatology of the subantarctic islands [*Aubert*, 1967]. In addition to making systematic observations on the flora and fauna, a number of conclusions regarding human meteoropathology were drawn.

The meteoropathological significance of strong and rapid variations in atmospheric pressure during the passage of rapidly moving depressions was repeatedly noted [*Pruche*, 1961]. Certain psychological and psychosomatic reactions appeared to be related to these meteorological phenomena. It is emphasized that considerable variations in weather conditions occur on Iles Crozet and Kerguelen within short periods of time or close together.

In addition to collecting general data on adaptation to temperature variations and to heat loss due in particular to the wind, more specialized studies were made of resistance to cold during diving in subantarctic waters where the temperature was about +5°C. In these studies the importance of solar radiation was noted, and adaptation seemed to be related more to psychological factors than to physiological factors [*Grua*, 1966, 1967; *Pruche*, 1961].

Pathological reactions are commonplace in the subantarctic islands, and the incidence of trauma is high. There were some cases of surgical emergencies. Psychopathology appears to be the predominant factor in morbidity on the subantarctic islands, particularly on Ile Amsterdam. Several possible factors appear to be involved: superficial selection of participants, lack of homogeneity of the population, relatively low group cohesion, and in particular the frequent atmospheric disturbances, the wind, and the oceanic climate. On Ile Amsterdam the geological or geographic features could be a factor: the island is a volcano emerging from the sea, absolutely circular in shape, and the inhabitants wherever they look can see only the immensity of the ocean surrounding them or the slopes of black lava pitted with craters [*Diviné*, 1960].

An epidemic of Queensland fever occurred during one winter on Ile Kerguelen [*Bourlaud*, 1957]. This outbreak was probably associated with the importation of sheep from Madagascar onto the island.

A psychophysiological investigation was conducted on Ile Amsterdam to determine the correlation between atmospheric pressure and the electrical conductivity of the skin. As a corollary to this concept it was hypothesized that there would be a negative correlation between emotional reactivity and atmospheric pressure. This hypothesis was tentatively confirmed in a related study of neuromuscular excitability [*Bessuges*, 1963, 1967, p. 884].

Current research. Research of the type now in progress on the Adélie Coast, which is in accordance with the recommendations of the Subcommittee on Human Adaptability Programs in the Antarctic, has been initiated also on the subantarctic islands. In addition to conducting systematic testing of all wintering subjects at fixed times, environmental measurements are being made, particularly on Ile Kerguelen. Psychological research is being emphasized on Ile Amsterdam.

BELGIAN STATION

The Belgian Roi Baudouin base is situated at 70°26′S, 24°19′E. The station was opened during 1958–1960; it was reoccupied during 1965–1966 by a joint Belgian and Dutch expedition.

As far as is known, the only published medical research is related to the 1959 expedition. Two monographs [*Staquet*, 1961*a*, *b*] were concerned with various aspects of polar medicine: pathology, hygiene, nutrition, clothing, psychopathology, and so on.

Caloric intake by groups of subjects living inside or outside the base was investigated by *Staquet* [1961*b*]. The subjects living on the base had very comfortable living conditions and were not unduly exposed to cold. On the average they spent 23 of 24 hours indoors. The mean caloric intake per person in this group was 3481 cal, that is, 48 cal per kilogram body weight per 24 hours.

The subjects living outside the base had to endure more severe exposure to cold because their tour of duty was during the depth of the polar winter and they had no heat supply. On the other hand, they had adequate protective equipment. Some were engaged in moderate physical activity, and others were engaged in more strenuous activity. Nevertheless, the mean daily intake of calories for these persons living outside the base was 3553 cal/day, that is, 52 cal per kilogram body weight per 24 hours. It was conlcuded that there was very little increase in the daily ration of food consumed by subjects living in the cold in comparison with that consumed by subjects well protected against the cold.

In a study of urinary excretion of 17-ketosteroids and 17-hydroxysteroids, subjects were examined on six occasions: before departure from Belgium, soon after arrival at Roi Baudouin base, after 3 months, during a traverse, during a survival study, and during a period of living under minimal shelter [*Staquet*, 1961*b*, 1962].

The following results were obtained. The excretion of steroids by persons living on base who were not unduly exposed to cold was normal. Persons exposed to low temperatures and not engaged in strenuous work but provided with adequate polar equipment and thus feeling cold only occasionally showed a decreased excretion of 17-ketosteroids and 17-hydroxysteroids. If in addition to these factors the subjects were deprived of food, the steroid excretion was reduced still further. Intensive physical work did not produce a significant fall in excretion. In some cases the diminution of urinary steroid excretion persisted for several days after the return to a temperate environment.

In summary, low temperatures produced a significant and prolonged decrease in the excretion of these metabolites. Metabolism of the adrenal hormones was modified, but it was impossible to say whether this modification was due to variation in the level of adrenal secretion or to changes in catabolism. This specific response to cold could be regarded as a component of acclimatization.

NORWEGIAN-BRITISH-SWEDISH STATION

The Norwegian-British-Swedish Antarctic Expedition wintered during 2 consecutive years (1949–1952) at Maudheim in Queen Maud Land. The first year the expedition comprised 15 men, and the second year it comprised 14 men. The station was located at 71°02′S, 10°55′W.

Numerous medical investigations were carried out by the expedition medical officer. In one study, blood samples were taken once a month, and the white cell and red cell counts, the differential white cell count, the hemoglobin concentration, the blood adrenalin concentration, and the sugar concentration were determined [*Wilson*, 1953].

The following results were reported. No change in the number of red cells and no significant seasonal variations were found. There was an increase in hemoglobin concentration in winter and spring and a slight decrease in summer.

The white cells were counted in the morning and the evening. The morning count was the most characteristic and the most stable; the cell count was low at the beginning of the stay at Maudheim but showed a significant rise during the winter. The investigator attributed this result to adaptation to a different climate.

There was no change in the differential white cell count and no change in the blood adrenalin. The blood sugar curve, showing a decrease in the first months and a maximum level in winter, was similar to that for white cells.

In another series of studies, basal metabolism (BMR) was related to climatic variations [*Wilson*, 1956*a*, *b*, 1962, 1966]. The outdoor temperature was frequently around -40°C and was accompanied by winds for long periods. The indoor temperature was about 14°C. Systematic measurements of BMR during the two winters revealed seasonal variations with maximums in May and September. The lowest values were obtained in June and August. The range between extremes of BMR was from 4.1 to 5.0%. There was no correlation with changes in body weight.

The variations in BMR were thought to be related to modifications in the rhythm of life and work rather than to climatic changes directly. Because no relationship between variations in BMR and changes in climate was apparent, there was no indication that acclimatization had taken place.

The respiratory rate, ranging between 10.7 and 12.2 per minute, remained relatively constant over all measurements made during the two winters [*Wilson,* 1956*b*]. The respiratory rate had no relationship to BMR. Similarly, no correlations with temperature variations or the level of ultraviolet radiation were observed.

Body weight of wintering personnel showed a net increase, amounting to about 2.7–5.3 kg, in the course of the two winters [*Wilson*, 1960]. Corresponding to this general increase in body weight, the weight curve rose during the winter and fell in the spring and summer. These weight increases were associated with periods of sedentary life. The mean loss of weight of the participants during traverses was 3 kg.

The climate of Maudheim, i.e., temperature, wind velocity, solar radiation, and wind chill, was studied [*Wilson*, 1957]. Siple's formula [*Siple and Passel*, 1945] for calculating wind chill was employed, and data were given to characterize local cooling power. Wind chill values were used to compare the climate of Maudheim with that of other antarctic stations. The wind chill formula showed no direct correlation with outdoor temperature, but there was a seasonal correlation with outdoor temperature. The importance of solar radiation and drift also was stressed.

During winterings at Maudheim, 69 cases of frostbite were recorded, and these cases were studied in relation to wind chill [*Wilson*, 1963, 1964]. Although they occurred at very different temperatures, ranging between -8°C and -46°C, it was noted that they all occurred at wind chill values between 1400 and 2100.

The probability of the development of frostbite was shown to depend on the diameter of the organ exposed or, more precisely, on its volume. The danger levels for wind chill values varied depending on the organ concerned.

Two important monographs describing medical research conducted at various polar bases and in particular at Maudheim were published in 1965 and 1966 [*Wilson*, 1965, 1966]. Climatic factors were considered, special reference being made to wind chill, radiation, absence of light, humidity, living quarters, clothing, food, variations in body weight and skin thickness, and, finally, pathology. In a chapter on physiological adaptation the arguments presented by various investigators were examined with respect to possible acclimatization to cold [*Wilson*, 1965, pp. 711–729]. Changes in BMR, respiration, body fat, serum lipids, blood adrenalin, blood sugar, adrenocortical activity, thyroid activity, diuresis and urinary excretion, diurnal rhythm, and thermal response were examined, and evidence for peripheral acclimatization was presented. Selection problems and psychological reactions of wintering personnel were also discussed.

JAPANESE STATION

The Japanese base of Shōwa (Syowa) is at 69°00′S, 39°35′E. It was first opened in 1957 and operated until February 1962. The station was reopened in 1966 and is still operating.

Most of the research from this station was published in Japanese. Only brief accounts of these investigations can be given here.

The research program envisaged for Japanese expeditions was described by *Ito* [1959] and *Ogata* [1959]. The program was concerned with physiological variations during the voyage to Antarctica, acclimatization during the winter period, and fluctuations in physiological rhythms during the 24-hour cycle.

The biometeorological environment was studied by recording the air temperature and measuring cooling effects by means of wet and dry bulb catathermometers [*Murakoshi and Morita*, 1963]. Ultraviolet radiation was measured by a photometer, and solar radiation was measured by a black bulb thermometer.

Systematic measurements of body weight, arterial pressure, central temperature, BMR, and blood counts were made, and urinary excretion of 17-ketosteroids and electrolytes was determined. The results were described by *Watanabe* [1967].

Body weight generally increased during the stay in Antarctica. Increases occurred mostly during the winter, slight decreases in body weight occurring during the summer.

Arterial blood pressure showed a slight decrease in the winter, probably associated with the routine mode of life.

Central temperature, varying with changes in climate and periods of seasickness during the voyage to Antarctica, declined slightly during the winter.

BMR during the voyage varied with outside temperature in a manner similar to the way that it varies according to seasonal variations in Japan, that is, increasing in the winter and decreasing in the summer. During the antarctic winter, BMR remained relatively constant and slightly below the mean normal level [*Azuma*, 1962].

The blood formula showed a constant red cell count, but the number of polymorphs fell slightly during the winter.

The urinary excretion of 17-ketosteroids was higher in the winter than in the autumn during the stay in Antarctica. Comparison with controls showed that the urinary excretion of sodium and potassium was higher in the antarctic wintering subject.

Fluctuations in the number of circulating blood eosinophils and in the urinary excretion of 17-ketosteroids in the 24-hour period, as well as fluctuations in the urinary elimination of sodium and potassium, were similar to those in the controls [*Muto et al.*, 1968]. However, variations were more irregular in the nonwintering subjects than in the wintering subjects.

One report was concerned with nutrition during the wintering period [*Hara*, 1964]. Altogether, more than 300 different foods were used during the winter to provide a sufficient variety of menus. The quantity of food ingested was about 3500 calories per person per day, including more than 180 grams protein per day. No nutritional problems were observed.

Some data on social and living conditions were provided by *Kitamura* [1963], *Matsuda* [1964], and *Muto* [1963] and on diseases and injuries by *Yoshioka* [1963]. For lack of information, it is not possible to provide an exhaustive account of Japanese research, and, regrettably, no information concerning recent or current Japanese research is available.

NORWEGIAN–SOUTH AFRICAN STATION

A Norwegian station, located at 70°30′S, 2°52′W, was opened in 1959 and occupied until 1963. Since 1960 it was operated by South African expeditions. Unfortunately, nothing is known of research that may have been carried out at this station.

REFERENCES

French Stations on the Adélie Coast

Artignan, P., and J.-C. Guériot, Hematological studies of subjects wintering at Dumont d'Urville base (Adélie Coast), *Ann. Hydrol. Climatol.*, *42*(109), 35–48, 1971.

Goy, G., The psychological climate of wintering, thesis in medicine, Univ. of Paris, Paris, 1959.

Nègre, C., Fatigue in the polar regions, thesis in medicine, Univ. of Paris, Paris, 1960.

Perrier, F. J., Psychopathology in polar expeditions, in *Symposium on Antarctic Biology*, p. 564, Hermann, Paris, 1964.

Perrier, F. J., On some problems presented by human homosexuality in polar expeditions, *Concours Med.*, *13*, 1–3, 1965.

Perrier, F. J., Psychopathology and the polar climate, in *Biometeorology*, vol. 2, part 2, edited by S. W. Tromp, pp. 846–851, Pergamon, New York, 1967.

Rivolier, J., Selection, protection and adaptation of subjects having to live in the cold, thesis in medicine, Univ. of Paris, Paris, 1953.

Rivolier, J., Some problems facing the medical officer of a polar expedition, *Concours Med.*, *12*, 2045, 2159, 1954.

Rivolier, J., Biological climatology of Géologie Archipelago (Adélie Coast), *Presse Therm. Clim.*, *5–6*, 135–142, 1955*a*.

Rivolier, J., Cold and altitude and their relationship to food supply, *Ann. Nutr. Aliment.*, *9*, 135–177, 1955*b*.

Rivolier, J., Accidents due to cold, in *Encyclopédie Médico-Chirurgicale*, *Agents Physiques*, vol. 16508 A10, pp. 1–10, Editions Techniques, Paris, 1957.

Rivolier, J., Some psychological aspects of problems of adaptation in wintering, *Doc. 15*, 13 pp., World Health Organ., Geneva, Switzerland, 1962*a*.

Rivolier, J., Preventive measures against the appearance and spread of new diseases in arctic and antarctic regions, *Doc. 17*, 8 pp., World Health Organ., Geneva, Switzerland, 1962*b*.

Rivolier, J., Two problems of medical logistics, in *Proceedings of the Symposium on Antarctic Logistics*, pp. 574–578, National Academy of Sciences–National Research Council, Washington, D. C., 1963*a*.

Rivolier, J., Rations of the International Glaciological Expedition to Greenland, in *Proceedings of the Symposium on Antarctic Logistics*, pp. 543–573, National Academy of Sciences–National Research Council, Washington, D. C., 1963*b*. (Reproduced in *Bull. Soc. Hyg. Aliment.*, *52*, 70–93, 1964.)

Rivolier, J., Medical surveillance and selection of members of French Polar Expeditions, *Concours Med.*, *21*, 3423–3429, 1964.

Rivolier, J., Measuring instruments and research possibilities in bioclimatology, in *Biometeorology*, vol. 2, part 2, edited by S. W. Tromp, pp. 805–813, Pergamon, New York, 1967*a*.

Rivolier, J., Nutrition and the polar climate, in *Biometeorology*, vol. 2, part 2, edited by S. W. Tromp, pp. 852–859, Pergamon, New York, 1967*b*.

Rivolier, J., and G. LeBideau, Carbohydrates and cold, *Bull. Soc. Hyg. Aliment.*, *48*, 11–23, 1960.

Rivolier, J., and F. J. Perrier, Psychosomatic problems and psychological adaptation to polar wintering, An attempt at biometeorological correlation, in *Publication of a Symposium in Human Adaptability and Its Methodology*, edited by H. Yoshimura and L. Weiner, Japanese Society for the Promotion of Sciences, Tokyo, 1966. (Reproduced in *Rev. Med. Psychosom.*, *2*, 1966.)

Rivolier, J., J.-C. Guériot, and F. J. Perrier, Study of fatigue and its physiopathological consequences in subjects wintering in the Antarctic, *Sem. Hop.*, *43*, 3510–3515, 1967.

Sapin-Jaloustre, J., Contribution to the study of acclimatization of man and the higher vertebrates in the Antarctic and, in particular, on the Adélie Coast, thesis in medicine, Univ. of Paris, Paris, 1955.

Sapin-Jaloustre, J., *An Investigation of Frostbite: Observations Made by the First French Antarctic Expedition on the Adélie Coast*, Hermann, Paris, 1956.

Siple, P. A., and Ch. F. Passel, Measurements of dry atmospheric cooling in subfreezing temperature, *Proc. Amer. Philos. Soc.*, *89*(1), 177–199, 1945.

Soussen, G., and R. Digo, Psychological problems of adaptation of small groups of people in an isolated situation in polar climates, report, Center for Flying Personnel, Air Min., Paris, 1964.

Stations on French Subantarctic Islands

Aubert, M., Contribution to subantarctic biometeorological studies: Kerguelen, Crozet, New Amsterdam, in *Biometeorology*, vol. 2, part 2, edited by S. W. Tromp, pp. 785–795, Pergamon, New York, 1967.

Bessuges, J., Psycho-physiological perspectives, *Com. Nat. Fr. Rech. Antarctiques, 6*, 1–140, 1963.

Bessuges, J., An attempt to investigate determinism in human behavior in an antarctic environment, in *Biometeorology*, vol. 2, part 2, edited by S. W. Tromp, Pergamon, New York, 1967.

Bourlaud, J., Pseudoinfluenzal symptoms of epidemic type observed on Kerguelen among Europeans during the subantarctic winter and spring: Resistance of Comoro Islanders, *Bull. Soc. Pathol. Exot.*, *50*, 870–874, 1957.

Diviné, Medicine in a closed vessel, *Bull. Territ. Terres Australes Antarctiques Fr.*, *11*, 14–31, 1960.

Grua, P., The psycho-physiological influence of sunshine, especially on resistance to cold in antarctic diving, paper presented at 4th International Biometeorological Congress, Rutgers Univ., New Brunswick, N. J., Aug. 26 to Sept. 2, 1966.

Grua, P., The fight against cold and rapid adaptation to diving conditions in cold water (Kerguelen), in *Biometeorology*, vol. 2, part 2, edited by S. W. Tromp, pp. 831–837, Pergamon, New York, 1967.

Pruche, Y., Adaptability to the climate and thermoregulation, *Bull. Territ. Terres Australes Antarctiques Fr.*, *17*, 4–20, 1961.

Belgian Station

Staquet, M., Observations médicales et physiologiques sur une expédition au Pole Sud, *Bruxelles Med.*, *41*, 741–751, 1961*a*.

Staquet, M., Besoins énergétiques et activité corticosurrénale chez les membres d'une expédition polaire, *Arch. Belg. Med. Soc. Hyg. Med. Trav. Med. Leg.*, *19*, 689–712, 1961*b*.

Staquet, M., Variation de l'élimination des stéroides dans l'urine en climat polaire, *Polar Rec.*, *11*, 317, 1962. (Reproduced in *Antarctic Biology: Proceedings of the First SCAR Symposium*, edited by R. Carrick, M. Holdgate, and J. Prevost, p. 515, Hermann, Paris, 1964.)

Norwegian-British-Swedish Station

Wilson, O., Physiological changes in blood in the Antarctic, A preliminary report, *Brit. Med. J.*, *3*, 1425–1428, 1953.

Wilson, O., Adaptation of the basal metabolic rate of man to climate, *Metabolism*, *5*, 531–542, 1956*a*.

Wilson, O., Basal metabolic rate in the Antarctic, *Metabolism*, *5*, 543–554, 1956*b*.

Wilson, O., Objective evaluation on wind chill index by records of frostbite in the Antarctic, *Int. J. Biometeorol.*, *11*, 29–32, 1957.

Wilson, O., Changes in body weight of men in the Antarctic, *Brit. J. Nutr.*, *14*, 391–401, 1960.

Wilson, O., Cooling effects of an antarctic climate on man, with some observations on the occurrence of frostbite, *Nor. Polarinst. Skr.*, *128*, 1963.

Wilson, O., Atmospheric cooling and the occurrence of frostbite in exposed skin, in *Symposium on Arctic Biology and Medicine*, vol. 4, *Frostbite*, edited by E. G. Viereck, pp. 73–86, Arctic Aeromedical Laboratory, Fort Wainwright, Alaska, 1964.

Wilson, O., Human adaptation to life in Antarctica, in *Biogeography and Ecology in Antarctica, Monogr. Biol.*, vol. 15, edited by P. van Oye, J. van Mieghem, and J. Schell, pp. 690–752, W. Junk, The Hague, Netherlands, 1965.

Wilson, O., Field studies on the effect of cold on man, with special reference to metabolic rate, *Acta Univ. Lund., Sect. 2, M. M.*, 1966.

Wilson, O., Basal metabolic rate of 'tropical' man in a polar climate, in *Biometeorology*, vol. 2, part 2, edited by S. W. Tromp, pp. 411–426, Pergamon, New York, 1967.

Japanese Station

Azuma, T., Acclimatization of physiological function in Antarctica, *Antarctic Rec.*, *17*, 69–70, 1962.

Hara, M., Report on the food provided for the Japanese Antarctic Research Expedition, *Sci. Rep. 1, Ser. F*, Jap. Antarctic Res. Exped., Tokyo, 1964.

Ito, Y., Report on activities on medical subcommittee and medical team for the Japanese Antarctic Research Expedition 1956-67, part 1, *Antarctic Rec.*, *6*, 54–72, 1959.

Kitamura, T., Human relation in wintering life in the Japanese Antarctic Research Expedition, *Antarctic Rec.*, *17*, 73–74, 1963.

Matsuda, T., Some notes on the social life of 16 persons of the Japanese antarctic wintering party, *Antarctic Rec.*, *20*, 75–87, 1964.

Murakoshi, N., and Y. Morita, Characteristics of the weather observed at Syowa station, *Antarctic Rec.*, *17*, 75–76, 1963.

Muto, A., Living conditions of antarctic wintering, *Antarctic Rec.*, *17*, 72–73, 1963.

Muto, A., S. Nakano, K. Suzuki, T. Sato, and T. Sakai, On the annual variation of serum protein fractions in the Antarctic, *Biochem. Phys.*, *13*, 91–95, 1968.

Ogata, M., Report on physiological results of the Japanese Antarctic Research Expedition, 1, *Antarctic Rec.*, *6*, 47–53, 1959.

Watanabe, K., The physiological acclimatization in the fourth Japanese Antarctic Research Expedition 1959-61, in *Biometeorology*, vol. 2, part 2, edited by S. W. Tromp, pp. 880–884, Pergamon, New York, 1967.

Yoshioka, T., Diseases and injuries in the Antarctic, *Antarctic Rec.*, *17*, 70–71, 1963.

PHYSIOLOGICAL RESEARCH AT U.S. STATIONS IN ANTARCTICA

JAY T. SHURLEY

Behavioral Sciences Laboratories, Veterans Administration Hospital, Oklahoma City, Oklahoma 73104
Department of Psychiatry and Behavioral Sciences, University of Oklahoma Health Sciences Center
Oklahoma City, Oklahoma 73190

Abstract. The biomedical and physiological research at U.S. antarctic stations published since the International Geophysical Year in 1956 is reviewed and summarized. Some previously unpublished data gathered by J.T. Shurley et al. from the Oklahoma sleep project (1966–1971) are included. Reports from this project form the bulk of the article. Detailed descriptions of the macroenvironment (South Polar Plateau) and the microenvironment (South Pole station) relevant to biology and medicine are given, together with a brief description of the study population (South Pole station personnel) for the 1967 and the 1968 austral winters. Findings of the core project on the psychophysiology of sleep and dreaming, EEG's under photic stimulation, sleep and activity patterns, the hematological and cardiopulmonary survey (documenting the hypoxic stress), erythropoietin level measurements in blood and urine, the acute effects of hypoxia on sleep psychophysiology, and the neutropenic response of station personnel to prolonged isolation at Plateau station (United States) are reviewed in detail. A summary statement of the individual clinical response to the antarctic wintering experience concludes that modern, sophisticated methods of biomedical research are now feasible and that their application to the many problems of human adaptation to the harsh antarctic environment portends a rich reward of increased knowledge of the ecology of the human species.

From the beginning of the International Geophysical Year (IGY) deployment in 1956 until the 5-year Oklahoma Medical Research Foundation sleep project entered the field in Antarctica in November 1966, human physiological and psychological studies at U.S. stations in Antarctica were brief, sporadic, and extremely limited in scope [*Sladen,* 1965]. Adequate reviews of the research performed during this period have previously been published by *Wilson* [1965], *Lewis and Masterton* [1963], *Edholm* [1965], *Gunderson* [1968], and the *Committee on Polar Research* [1970]; therefore only occasional reference is made to the work covered by these reviews.

The research picture has remained generally unchanged since then, although a small number of medical officers from the U.S. Navy Task Force have conducted studies on wintering-over personnel at each U.S. station for the past several years, and it is expected that their data will be published in due course. Aside from the Oklahoma project, however, only a handful of American reports (contributed by U.S. Navy medical officers) has appeared in the literature. *Taylor* [1960], as medical officer at McMurdo station in 1956–1957, reported his IGY experiences. *Beazley* [1968] reported his medical experiences at South Pole station during the 1964–1965 season. He observed mild altitude sickness in new arrivals, elevated hematocrits within 30 days, and mild cases of tracheobronchitis attributed to inspiration of the cold air. *Hunt* [1968] described his clinical observations on the station complement at Byrd station during the 1966 winter and noted elevations in both hematocrit and white cell count. Most men lost weight, but no man showed hypertensive changes, this observation being in apparent contrast to observations in other years at this station.

Pratt et al. [1966] described an ambitious program to evaluate the base line and acute cardiorespiratory acclimatization data of men selected to spend the winter at Plateau station; the project was designed as a preliminary step forward in identifying men who might later develop pulmonary edema on ascent to high elevations. Before deployment, when his eight subjects were exposed to an altitude chamber test to 4300 meters (14,000 feet) for 36 hours, two suffered protracted nausea and vomiting and had to be removed prematurely. After arrival at Plateau station (elevation 3600 meters, or 11,892 feet), however, no man developed pulmonary edema.

In the next year at the same station, *Blackburn*

[1968] described his findings of decreased white cell counts, delayed rise in hemoglobin concentration, increased lung volumes, and no apparent electrocardiographic changes. He collected additional physiological and psychological data bearing on acclimatization and adaptation to this most isolated antarctic station, but the results were not reported in this account.

Finally, *Tyler* [1968], who served as task force medical officer for the U.S. Navy Task Force 43 (Operation Deep Freeze), discussed the logistical aspects of medical activities conducted for the U.S. Antarctic Research Program (USARP) including personnel screening, selection of medical support personnel, station medical facilities, organization of services, and medical evacuation procedures. Ten percent of all sick call visits to all stations were due to upper respiratory infections and minor accidental injuries. Cold injuries were very rare, and minor complaints of ear infections, eye injuries, dermatological problems, and gastrointestinal illnesses were found to occur in about the same incidence as they occur in medical practice elsewhere. Plans for a coordinated medical research program to be carried out by station medical officers were outlined in Tyler's report; reports of the findings are anticipated.

The present review is concerned primarily with publications emanating from the Oklahoma sleep project. Since this was the first research in human biology to be conducted under USARP auspices, the principal investigators had no previous experience in antarctic fieldwork and were necessarily somewhat naive about field conditions there. (However, one of the principal investigators, Chester M. Pierce, had made a month's visit to U.S. stations in Antarctica in November 1963 and thus was able to form some valuable impressions of working conditions there before the proposal was submitted.) The way in which they mastered the many necessary lessons and the first fruits of their efforts are described in the April 1970 issue of the *Archives of Internal Medicine,* devoted to a symposium entitled Man on the South Polar Plateau, for which *Shurley* [1970*b*] served as guest editor. This journal issue contains a number of articles constituting the major published record to date of the Oklahoma sleep project and is the principal source for most of the reviews below.

The subjects studied in these investigations were unpaid volunteers serving as scientist-technicians and Naval Support Force personnel assigned to South Pole and Plateau stations. Despite information from many sources that such personnel commonly reject or offer minimal cooperation to psychologically oriented researchers, no effort was made to conceal the identity of the psychiatrists and psychologists on the research team or the psychological aspects of the proposed studies.

Since nothing was known of the psychophysiological responses of humans during sleep to the multiple stresses of antarctic life through the annual cycle, this study was designed primarily to document what, if any, quantitative changes characterize the electrophysiological patterns of sleeping and dreaming in wintering-over personnel. Such changes were to be correlated with measurable alterations in the physical and the social aspects of station life during the 12-month period of residence at South Pole station. A secondary aim was to obtain a more integrated picture of the psychophysiology of men adapting to the unique environment of the South Polar Plateau and to evaluate the feasibility of using modern laboratory methods to study these processes in the field. A third aim was to accumulate observations in this real life field situation and use these to construct appropriate hypotheses concerning the behavior of humans in an isolated hypodynamic environment.

CHARACTERISTICS AND MAKEUP OF THE STUDY POPULATIONS

Eleven men from the 21 members of the 1967 wintering-over party and nine men from the 19 members of the 1968 party volunteered for the sleep studies at South Pole station. All 20 subjects were Caucasian males. In 1967 the mean age was 28.2 years (range of 23–35 years), and in 1968 it was 30.0 years (range of 23–49 years). All volunteers except one who had spent the winter of 1966 at McMurdo station had no previous experience in Antarctica. None of the men had previously been acquainted with any of the others. Fourteen were single, and six were married and had been living with their wives before departing for Antarctica. Nine of the 20 were in naval service, and 11 were civilians.

The mean educational level of all subjects was 14 years (range of 8–21 years). By occupational groups, 11 served in scientific-technical positions, four were Navy administrative-technical (white-collar) personnel, and five were skilled craftsmen in the Navy Seabee (blue-collar) group. Each man served as a specialist in one of the following occupational categories: medical officer, officer in charge, weather observer, cosmic ray observer, geomagnetic and seismology observer, pharmacist's mate, biomedical technician, psychophysiology technician, radioman, ionospheric and radio wave propagation observer, construction electrician, electronic technician, and builder-carpenter.

All volunteers were healthy, having passed rigid Navy physical and mental qualification examinations for antarctic duty.

SALIENT CHARACTERISTICS OF THE MACROENVIRONMENT

The U.S. Amundsen-Scott South Pole station is at the southern spin axis of the earth, at 90°S. It rests on the vast ice sheet of the South Polar Plateau, which averages over 2500 meters (8000 feet) thick and rises 2800 meters (9200 feet) above sea level. The mean annual barometric pressure is around 500 mm Hg (680 millibars), qualifying it as a high-altitude observatory, and the mean pressure altitude is equivalent to approximately 3350 meters (11,000 feet).

Located within the globe's greatest heat sink, it is one of the coldest spots on the earth's surface. When the Oklahoma biomedical project was in progress, the maximum temperature was a record-breaking -1°F (-18°C) in 1967 and -9°F (-22.5°C) in 1968. The lowest temperatures were -105°F (-76.1°C) in 1967 and -112°F (-80°C) in 1968. For comparison, the lowest surface temperature registered in Antarctica (and therefore on the globe) in any year was -126.9°F (-88.3°C) at Vostok station (USSR) near the south geomagnetic pole. When the wind chill factor is taken into consideration, even at a minimum wind speed of 1 km/hr (1 mi/hr) and average temperatures, the environment is bitterly cold. At the average annual temperature (-60°F) and wind velocity, wind chill figures indicate that exposed flesh freezes in 30 sec or less.

The South Polar Plateau is also extremely dry. Indeed, the relative humidity is too low to be measured accurately by available methods. Its mean annual precipitation (as snow) of less than 1.2 cm (½ inch) makes the South Polar Plateau the world's driest desert.

Night and day in the temperate zone sense are absent. There is one continuous day (summer) of 6 months' duration and one continuous night (winter) of equal duration. The annual sunrise occurs on September 22, and sunset occurs on March 22, about a 2-week period of twilight occurring around each. Clock time is arbitrarily set to coincide with time at McMurdo station and Christchurch, New Zealand. Even the day of the week is arbitrarily selected from one of two choices, since South Pole station straddles the international date line.

Because of the severe weather conditions, neither surface travel nor air travel is considered technically possible except between about October 25 and February 25 each year. Therefore geographic isolation is absolute for approximately 8 months each year. This isolation more than any other feature attracted the interest of the author, whose researches previously involved the effects of sensory, perceptual, and social isolation on intact, healthy humans.

SALIENT CHARACTERISTICS OF THE MICROENVIRONMENT OF SOUTH POLE STATION

At the time of the study, South Pole station (Figures 1 and 2) consisted of the IGY buildings of 1956–1957, originally constructed on the snow surface but now drifted over with a snow cover averaging 8 meters (25 feet) in depth. Each man had to walk at least 12 meters (40 feet) through the tunnels between buildings several times each day for sleeping, eating, using toilet and washroom facilities, and getting to specific work sites. Two work sites (cosmic ray and riometer) were nearly 400 meters (¼ mile) from the main camp and could be reached only by walking over the snow along pathways marked by poles and rope. Each man had his own private sleeping quarters with a maximum floor space of approximately 5 m^2 (48 ft^2). Although the heating system kept the air 2 meters (5–6 feet) above the floor at 70°–80°F, the layering effect was so pronounced that the floor-ceiling gradient consistently ranged from below 0°F to 85°F. Meals were taken in the galley, where the relative humidity often exceeded 80%, in extreme contrast to the low humidty elsewhere.

All supplies and personnel were flown-in the 1320-km (820 miles) distance from Williams Field, near McMurdo station, after reaching Antarctica. Because imported diesel fuel, arctic, was used as the heat source to melt snow for the sole water supply, drinking water soon came to taste continually of diesel fuel. Moreover, the constant roar of the diesel-electric generators, the sole energy source of the station, permeated the buildings and surrounding area for at least a 400-meter distance, so that one had to make a special effort to escape it or else learn to live with the ear-splitting racket. The plume of exhaust smoke from the generators dissipated slowly into the constant wind. Because of the closely confined spaces in the buildings, one was constantly aware also of tobacco smoke. Here, of all places, just as in any urban industrial area, air and noise pollution were omnipresent.

A balanced daily ration of approximately 4000 calories/man was provided from long-stored frozen food stores, supplemented by fresh fruits and vegetables imported from New Zealand in the summer and by multivitamin capsules in the winter. (This allowance excludes the calories in alcoholic beverages, periodically consumed in considerable quantity.)

Fig. 1. Amundsen-Scott South Pole station in November 1967 at the supply tunnel entrance. The IGY buildings are now under 8–10 meters (10–15 feet) of snow cover. Note the air pollution from the diesel-fueled generators near the dome of the radiosonde tracking system.

Special lightweight clothing of cotton, wool, and synthetic fiber, weighing about 55 kg/man, or 25 lb, and designed for extreme dry cold, was worn regularly outdoors. A little less clothing was customarily worn in the tunnels, where no wind was present and the temperature averaged −20°F in the summer (open condition) and +10°F in the winter (closed condition). Because simple cold weather precautions were somewhat indifferently observed, minor frostbite of the cheeks and fingers was not uncommon. All personnel experienced a continuous watery nasal discharge and a frequent dry, hacking morning cough (some with blood-tinged sputum) throughout the year. The men experienced minor to severe shortness of breath, with or without exertion, on arrival. Dyspnea never completely subsided, even after a year in residence.

The intense cold rendered the outside air bacteriologically sterile. Consequently, upper respiratory tract infections virtually ceased within 2 weeks after the departure of the last plane in February and did not resume until the next plane arrived in November. However, molds, fungi, and their spores appeared to contaminate all food and clothing badly and appeared in all inhabited spaces. Cracks in the hands, lips, and nasal openings occurred frequently from the dryness, and small wounds healed very slowly, minor crusting and pus formation often being observed, but pyogenic skin infections common to other environments were not seen.

CORE PROJECT: PSYCHOPHYSIOLOGY OF SLEEP AND DREAMING

The basic aim of this project was to compare in detail the quantitative and qualitative aspects of the psychophysiology of sleeping and dreaming under two environmental conditions: the low-elevation temperate zone of urban industrial United States and the high-elevation polar zone of the isolated rural South Pole station. A secondary aim was to collect physiological, psychological, and social data that could possibly be correlated with the sleep and dream data. Finally, the project would help determine the feasibility of performing such sophisticated studies (ordinarily performed only in urban medical science centers) under remote field laboratory conditions.

On three consecutive nights the sleep of each volunteer was monitored according to the general method of *Aserinsky and Kleitman* [1953] and *Dement and Kleitman* [1957], using seven channels of electrophysiolog-

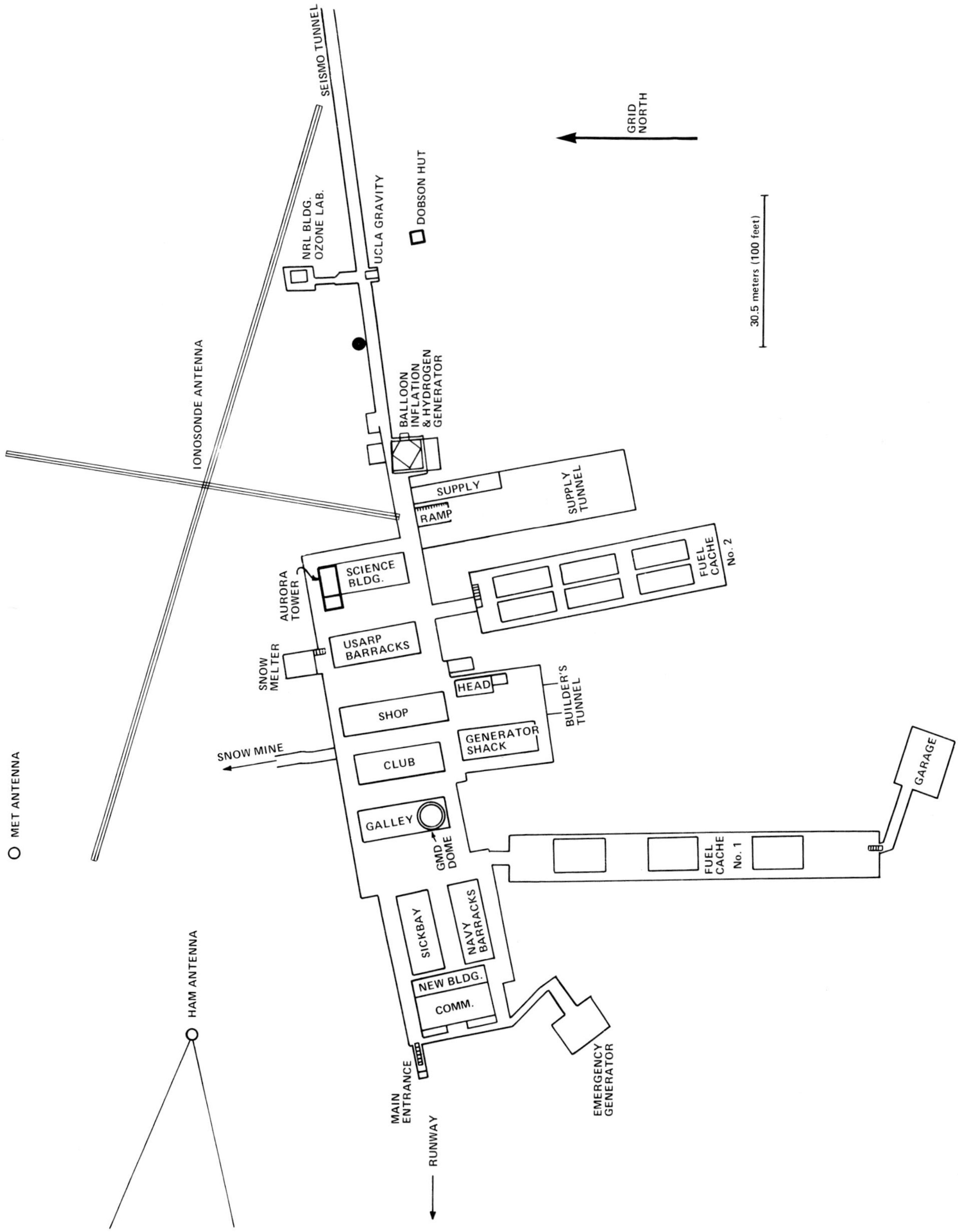

Fig. 2. South Pole station layout as of November 1968, showing the general arrangements of living, working, support, and scientific spaces.

TABLE 1. Signal Acquisition Array

Channel	Designation	Derivation	Time Constants, seconds
1	EOG	Left and right outer canthi bipolar	1.0
2	EEG	Left parietal and left occipital bipolar	0.3
3	EEG	Left frontal and left parietal bipolar	0.3
4	EMG	Left and right hyoid region	0.03
5	EEG	Right frontal and right parietal bipolar	0.3
6	EKG	Left and right arms	0.03
7	EDR	Left third finger, palmar surface	1.0

ical signals continuously recorded by polygraph in analog form on magnetic tape (Table 1).

The initial set of base line recordings of three nights of sleep was made in a two-room suite on an unused ward at the Quonset Naval Hospital, Rhode Island, before deployment of the volunteers to Antarctica. A cubby-hole in the science building and an adjoining tiny store-room were made available to the project at South Pole station in November 1966. After certain technical obstacles were overcome [*Brooks et al.*, 1973], satisfactory recordings were achieved, and four sets of data were recorded on each subject, one during each quarter of the year. Follow-up recordings were then obtained in the sleep laboratory of the Behavioral Sciences Laboratories, Veterans Administration Hospital, Oklahoma City, Oklahoma, within 60–360 days after the volunteers returned to the United States. Some self-report data were obtained concomitantly with these recording sessions. Thus for each subject a complete set of data amounted to 18 nights of sleep in six sets of three consecutive nights, along with certain auxiliary observations. For the entire project, 360 nights of sleep physiology were recorded from 20 subjects over a span of time from August 1966 through December 1969, comprising two consecutive antarctic fieldwork years, i.e., two polar days and nights (Table 2). Each night (8 hours) of sleep thus produced six or seven channels of physiological information, eventually displayed on approximately 430 meters (1400 feet) of strip chart. This provided detailed physiological information on central nervous system and associated bodily system responses during the resting and sleeping phases of the subjects' diurnal cycle.

The 475-kg (800 lb) Beckman instrument constituting the core of our data acquisition system (pictured in Figure 3 and described in detail by *Brooks et al.* [1973]) functioned very well indeed under probably the most hostile ambient conditions on the globe, conditions for which it was not of course designed. The tape-recorded analog records were returned by ship and plane to the base laboratory (Behavioral Sciences Laboratories) in Oklahoma City, where they were written out on paper strip charts and scored visuomanually according to the method described by *Rechtschaffen and Kales* [1968]. A corps of eight trained student and research assistant sleep scorers became highly competent in analog-to-digital conversion (sleep stage scoring) of the sleep data with an accuracy that no computer can yet match (Table 3). (Some of the techniques developed in this project for acquiring the highest quality of noise-free electroencephalographic signal recordings of human sleep have since been copied worldwide in sleep research laboratories.) These techniques represent a proven model for data acquisition under other exotic field conditions.

The formidable task of thoroughly analyzing these data in detail is still in progress. Results have exceeded all expectations in both quantity and quality and have saturated our modest data analysis capability. However, they have proved the feasibility of obtaining sophisticated biomedical data under extremely adverse field conditions.

TABLE 2. Data Acquisition Design

Period	Modal Dates	Code Designation	Location and Season
1	August 1966	TZ1	Temperate zone, Quonset Point, Rhode Island, summer
2	December 1966	SPS	South Pole station, summer
3	March 1967	W1	South Pole station, sunset
4	June 1967	W2	South Pole station, midwinter
5	September 1967	W3	South Pole station, sunrise
6	April 1968	TZ2	Temperate zone, Oklahoma City, Oklahoma, spring

The mean elapsed time from the first to the last recording was 19 months; the range was 18–22 months.

Fig. 3. Beckman biomedical recording and display system in use at South Pole station for acquisition of all-night sleep EEG and other electrophysiological data on 1.27-cm (½ inch) magnetic tape using the Ampex model SP-300 recorder/reproducer. The instrument is reconfigured specifically for this purpose and is operated inside a radio frequency shielding enclosure. K. Natani, psychophysiology technician and research assistant for the Oklahoma sleep project, USARP 1967, monitors the oscilloscope trace to check the operation of the system and observe changes in the sleeping subjects' levels of consciousness.

Our preliminary findings were reported by *Natani et al.* [1970] as long-term changes in sleep patterns of men on the South Polar Plateau. The data have clearly answered one question by showing that over this extreme range of environmental variation there was a negligible difference in the mean duration of sleep per 24 hours. For a sample of four men the mean sleep per 24 hours at the south pole was 448 min, whereas it was 463 min in the United States (Table 4). These EEG-verified average sleep durations agree closely with our own self-reported findings [*Shurley et al.*, 1970] of 453 min for 32 men at South Pole station during the summer and with those self-reported findings of *Lewis* [1961] of 474 min for all crew members of the 1952–1954 British North Greenland Expedition.

However, Natani has demonstrated that the mean time required to fall asleep after the subjects entered bed was longer among antarctic volunteers (15.4 min) than among the same men in temperate zones (3.5 min) (Table 5). Still, both figures are well within the 20-min latency for healthy sleepers as measured in U.S. laboratories [*Webb,* 1969]. Natani's finding thus fails to confirm the occurrence of the 'polar big eye' phenomenon (insomnia), reported by *Mullin* [1960], among men living at high latitudes for extended periods.

On the basis of the hypothesis of *Newman and Evans* [1965] and *Dewan* [1968] that the quantity of rapid eye movement (REM) sleep varies directly with information input load since the last period of sleep, we predicted that REM sleep would decline among men at the south pole as a result of the monotony of the environment. Indeed, antarctic volunteers had about one-fifth less REM sleep than they had before leaving the United States (27.9% of total sleep in the United States versus 23.9% in the Antarctic), but these figures failed to achieve statistical significance at the $p < 0.05$ level (Table 6).

The major and most surprising finding, however, was the virtually complete loss of stage 4 (slow-wave) sleep (SWS) at the end of the south polar winter. Moreover, SWS failed to return 6 months after the subjects' return to residence in the United States. From a base line mean of 20 min of stage 4, we found only 2.8 min in the winter 3 (W3) recordings and only 1 min after return to the United States (Table 7). Although the biological function of stage 4 sleep is unknown at present, we can speculate several interpretations of the data. One could speculate that neither stage 4 nor any other stage is biologically necessary for health, or it is possible that the

TABLE 3. Number and Percent of 20-Second Epochs on Which Raters Disagreed after the First Independent Reading

	TZ1	SPS	W1	W2	W3	TZ2
			Subject A305			
Total number of epochs	1320	1450	1381	1415	1347	1242
Number of disagreements	91	215	134	238	112	339
Disagreement, %	6.9	14.8	9.7	16.8	8.3	27.3
			Subject A310			
Total number of epochs	1166	1209	1404	1418	1228	1520
Number of disagreements	124	195	241	96	177	91
Disagreement, %	10.6	16.1	17.2	6.8	14.4	6.0

TABLE 4. Total Sleep Time

	Sequence	Sleep Time, min/24 hr: Subject 305	Subject 306	Subject 308	Subject 310	Mean
Temperate zone 1 (before base line)	1	439	437	...	390	422
South Pole summer	2	482	422	459	397	440
Winter 1	3	453	479	473	426	458
Winter 2	4	465	452	443	463	456
Winter 3	5	443	491	411	406	438
Temperate zone 2 (after base line)	6	413	496	534	502	486
Mean sleep time, min/24 hr		449	463	464	431	

loss of stage 4 represents the gradual depletion of some biochemical factor during chronic stress (e.g., hypoxia), along the lines of *Selye*'s [1950] postulated general adaptation syndrome. Alternatively, the absence of sunlight might somehow depress hypothalamic function, which in turn would diminish the pituitary secretion of the human growth hormone, whose release is known to be associated with the onset of stage 4 sleep [*Sassin et al.*, 1969]. These two alternative processes might be either irreversible or only slowly reversible. It is interesting to note that the selective deprivation of stage 4 sleep in healthy volunteers has been found to be associated with vague physical complaints, changes in bodily sensations, increased passivity and social withdrawal, and mild depression [*Agnew et al.*, 1964], symptoms quite similar to those reported by polar isolates at the end of the winter [*Gunderson,* 1963]. Apparently then, these silent changes in sleep physiology are related to subtle changes in sleep patterns, physical feelings, performance, mood, and affect, and they merit further investigation.

When each subject was awakened and asked for dream reports shortly after the beginning of each of two REM sleep periods on the third night of the recording period, the results were somewhat disappointing. Few dream reports were long enough or clear enough for content analysis by some of the standard methods currently in use [*Hall and Van De Castle,* 1966; *Gottschalk and Gleser,* 1969]. However, the above methods have been extended to allow statistically valid generalizations from shorter dreams, this development giving some hope that these data may yet be valuable in other than a subjective, purely clinical context. That dreams related by antarctic personnel have special significance and become long-remembered events in the life of the dreamer is well attested in *Shurley*'s [1970*a*] review of the dreams of Scott's northern party, reported by *Priestley* [1969] nearly 60 years later. Priestley's gripping first-person account of the ordeal of the party of six, who were unintentionally abandoned without supplies and forced to spend the 7 winter months in a hole in a snowbank on Inexpressible Island, is an epic of human survival under antarctic conditions. *Shurley*'s [1970*a*] speculated explanations for their survival include the production of vivid dreams about sumptuous banquets always denied in fantasy as they were in reality and the perfect self-discipline and mutual respect previously developed by

TABLE 5. Sleep Onset Latency

	Sequence	Sleep Onset Latency, min: Subject 305	Subject 306	Subject 308	Subject 310	Mean
Temperate zone 1 (before base line)	1	1	7	...	0	2.7
South Pole summer	2	2	38	15	7	15.5
Winter 1	3	7	13	4	2	6.5
Winter 2	4	7	19	50	10	21.5
Winter 3	5	4	5	61	3	18.3
Temperate zone 2 (after base line)	6	2	9	1	5	4.3
Mean sleep onset latency, min		3.8	15.2	26.2	4.5	

TABLE 6. Stage REM as a Percent of Total Sleep Time

		REM Sleep, %				
	Sequence	Subject 305	Subject 306	Subject 308	Subject 310	Mean
Temperate zone 1 (B)	1	30.7	27.2	...	26.0	27.96
South Pole summer	2	31.5	26.9	28.2	22.4	27.25
Winter 1	3	23.6	19.4	17.7	25.3	21.50
Winter 2	4	25.1	19.9	20.6	28.0	23.40
Winter 3	5	21.7	29.4	25.6	17.0	23.42
Temperate zone 2 (A)	6	20.7	20.5	26.6	20.7	22.12
Mean REM stage, %		25.6	23.9	23.7	23.2	

The mean percent stage REM for seven nights, temperate zones 1 and 2, was 24.62%. The mean percent stage REM for 16 nights, South Pole summer and winter 1, 2, and 3, was 23.89%.

the group in the field despite (or perhaps because of) their complete mutual self-revelation.

An overall concept of the detailed nature and extensive quantitative picture of sleep psychophysiology in relation to environmental change is conveyed by Tables 8–10. Table 8 contains the complete data summary for one subject for six nights of sleep. A similar table was constructed for each of the 20 subjects, comparisons thus being permitted within subject across time and geographical boundaries for each discriminable aspect of sleep patterning possibly related to adaptation. It illustrates well the stability of the mean sleep duration across conditions, the increase in sleep latency (insomnia) in three of four south polar recordings, the decreased percentage of REM sleep under early and midwinter polar conditions, the virtual loss of stage 4 sleep beyond base line and the failure to return to base line value on return to the United States, and the relative stability of all other measures across conditions. Yet all these fluctuations in sleep parameters would have to be considered fluctuations within the normal range for a healthy person and could in no way be considered indications of pathological sleep.

Tables 9 and 10 contain consolidated data summaries of mean values for all subjects' second nights of sleep across all conditions for the 1967 and 1968 seasons at South Pole station, together with grand means and standard deviations. Again, a picture of normal sleep, with impressive stability across all time periods and both years in most parameters, is presented. Yet the pattern of variations between the 2 years is itself rather variable and leads to speculation that the highly encoded information must somehow reflect that the essential subjective and objective experiences of all personnel were quite different and were perceived as being different between the 2 years.

Serafetinides et al. [1971] reported a study on new arrivals at South Pole station in which the possible stress of the acclimatization period was evaluated. Resting EEG's were made, and the effects of photic stimulation were measured on eight volunteers at South Pole station within their first week of arrival and again after an interval of 2 weeks. They found no EEG evidence of brain stress or impairment of gross brain reactivity during the initial acclimatization period and no evidence of tension or anxiety. On the contrary, the men tended to be inordinately drowsy, this drowsiness suggesting a readiness for relaxation. However, subtle brain impairment due to hypoxic or other stress could not be ruled out. These investigators concluded that studies should be designed

TABLE 7. Stage 4 Sleep as a Percent of Total Sleep for USARP 1967

		Stage 4 Sleep, %				
	Sequence	Subject 305	Subject 306	Subject 308	Subject 310	Mean
Temperate zone 1	1	2.7	10.5	...	0.3	4.5
South Pole summer	2	2.6	0.1	7.3	0.1	2.5
Winter 1	3	0.1	0.1	7.7	0.9	2.2
Winter 2	4	1.1	1.6	9.9	Ø	3.1
Winter 3	5	1.1	0.3	1.1	Ø	0.6
Temperate zone 2	6	0.8	0.3	Ø	0.1	0.3
Mean stage 4 sleep, %		1.4	2.2	5.2	0.2	

The usual values for the subjects in this study (males aged 20–35 years) were from 3 to 15%. Note the near disappearance of this sleep stage in all winter 3 and temperate zone 2 recordings.

TABLE 8. Data for Subject 306, Age 32

	TZ1	SPS	W1	W2	W3	TZ2	Mean
Date	Sept. 1966	Dec. 1966	March 1967	June 1967	Sept. 1967	March 1968	
Total sleep time, minutes	437	422	479	452	491	496	462.8
Sleep onset latency, minutes	7	38	13	19	5	9	15.2
REM onset latency, minutes	84	88	107	54	72	141	91.0
Number of awakenings	2	4	9	6	4	4	4.8
Number of stage changes	204	136	161	145	133	133	152.0
Number of movement arousals	77	110	72	109	123	83	95.7
Stage W (0), %	3.1	10.5	7.4	7.7	3.1	13.8	7.6
Stage M, %	1.0	0.5	1.1	0.4	0.1	0.4	0.6
Stage 1, %	7.6	9.0	11.7	8.2	7.3	5.7	8.2
Stage 2, %	42.2	58.8	60.0	57.5	55.6	56.7	55.1
Stage 3, %	9.7	3.1	3.1	9.6	5.8	4.4	5.9
Stage 4, %	10.5	0.1	0.1	1.6	0.3	0.3	2.1
Stage REM, %	27.2	26.9	19.4	19.9	29.4	20.5	23.9
Duration REM1, minutes	18	24	27	4	19	26	19.7
Duration REM2, minutes	19	26	17	18	38	25	23.8
Duration REM3, minutes	36	51	10	15	29	11	25.3
Duration REM4, minutes	60	21	29	53	7	43	35.5
Duration REM5, minutes	...	...	20	16	72	...	18.0
Total REM, minutes	119	114	93	90	145	102	110.5
Total NREM, minutes	318	308	386	362	347	395	352.7
Total Stage 4, minutes	46	0.3	0.3	7	1	1	9.3

to establish the roles of the four most probable factors involved in producing the unexpected drowsiness evidenced on the EEG records: mild chronic sleep deprivation, hypobaric hypoxia, low ambient temperature, and fatigue.

Shurley et al. [1970] reported the findings of a preliminary survey of the sleep and activity patterns of South Pole station personnel during the summer month December by using a daily self-report card of their own design (see *Natani and Shurley* [1974, Figure 1]). During a brief sample period (7 or 8 days), 32 of 34 volunteers contributed complete sets of data on time devoted to specific activities, showed high cooperation and reliability in reporting all data, and required very little instruction beyond the explanation printed on the cards. Not only were the hours of sleep (7.55 hours per 24 hours) consistent with those reported in similar surveys [*Kleitman and Kleitman,* 1953; *Lewis,* 1961; *Lewis and Masterton,* 1957], but the general patterns of sleeping and waking closely resembled those of men in urban settings. Of the time allocated to work, 47% was spent indoors alone, and despite one full holiday the mean work week was nearly 60 hours, the range being 18.5–101.5 hours. More than half the time spent in the limited recreational pursuits available (talking, games, reading, movies, and ham radio) was devoted to social conversation. In fact, social conversation accounted for a mean of 4.53 hr/man/day. In an analogous study of urban industrial workers (married men living at home), only half as much time was spent in social talk (mean of 2.27 hr/man/day) (W. L. Fesler and J. T. Shurley, unpublished data, 1967). This unexpectedly high output of social conversation in the party at South Pole station may partly reflect a compensatory response to the multiple novel stresses imposed by the setting, including the geographic isolation.

CARDIOPULMONARY STUDIES

Entering the field in 1966, we naively believed that the 2800-meter (9200 feet) elevation above sea level at South Pole station indicated a corresponding pressure altitude by temperate zone standards. Consequently, we were surprised to discover the severity of mountain sickness symptoms among new arrivals. We soon learned that a great disparity exists between the elevation and the recorded barometric pressure (range of 487–525 mm Hg), so that the actual pressure altitude falls in the 3350- to 3500-meter (11,000–11,500 feet) range throughout the year. This hypobaric, hypoxic environment falls within the critical range for human physiological functions influenced by hypoxia. As a result, with the collaboration of C. A. Guenter, we designed a study of the cardiorespiratory function of men at South Pole station, to be carried out concomitantly with our study of sleep psychophysiology. To our knowledge these are the first data published on the cardiorespiratory responses of temperate zone man living at high elevation for longer than 6 months.

Base line measures of cardiopulmonary, hematological, and metabolic functions were taken before the volunteers left the United States and at intervals over the year in 13 wintering-over subjects.

Guenter et al. [1970] have reported on the spirometry

TABLE 9. Data Summary for USARP 1967 South Pole Sleep for 11 Subjects

	TZ1 $N = 6$	SPS $N = 11$	W1 $N = 11$	W2 $N = 11$	W3 $N = 11$	TZ2 $N = 11$	Mean $N = 61$	Standard Deviation
Total record time, minutes	398.7	460.0	467.2	465.9	474.9	488.8	465.1	21.0
Total sleep time, minutes	391.5	448.3	450.2	445.9	457.8	478.5	445.4	28.9
Sleep onset latency, minutes	6.7	11.2	12.9	17.7	14.8	9.2	12.1	3.9
REM onset latency, minutes	91.0	82.3	76.2	86.7	69.3	116.6	87.0	16.4
Number of awakenings	2.7	5.0	6.1	8.1	6.7	4.1	5.5	2.0
Number of stage changes	153.2	164.5	161.1	160.4	144.5	150.1	155.6	7.6
Number of movement arousals	62.5	118.5	143.3	125.6	141.5	91.2	113.8	31.4
Total stage M, minutes	3.1	2.6	3.5	3.4	2.5	5.4	3.5	2.0
Total stage W (0), minutes	3.4	5.7	27.7	13.8	18.3	15.6	15.3	16.2
Total stage REM, minutes	97.2	104.1	88.0	93.9	102.0	106.0	98.5	6.8
Total stage 4, minutes	10.3	7.0	7.1	10.4	6.0	4.7	7.6	2.3
Total stage 3, minutes	31.3	27.9	22.7	26.4	25.8	22.6	26.1	3.3
Total stage 2, minutes	217.9	251.9	246.4	248.5	247.8	275.5	250.3	25.1
Total stage 1, minutes	27.3	47.0	54.6	48.6	55.0	47.2	49.2	29.1
Total stage NREM, minutes	287.9	335.8	331.8	334.8	335.6	345.7	328.6	20.5
Number of REM periods	3.5	4.3	4.2	4.4	4.5	4.2	4.2	0.6
Mean inter-REM onset interval, minutes	93.4	93.4	94.5	89.8	92.8	93.1	92.2	8.9
Mean inter-REM interval, minutes	70.0	70.3	72.3	69.6	71.0	68.6	70.2	8.4
Stage REM, %	23.8	23.1	19.7	20.9	22.3	22.1	22.0	1.5
Slow-wave sleep (stages 3 and 4), %	10.1	7.6	6.5	8.5	7.3	5.7	7.6	1.5
Stage 4, %	2.4	1.5	1.5	2.3	1.3	0.3	1.6	0.76
Stage 3, %	7.7	6.1	5.0	6.2	5.8	4.2	5.8	1.2
Stage 2, %	56.7	56.7	54.7	55.8	54.4	57.4	56.0	1.2
Stage 1, %	7.0	10.5	12.2	10.8	11.7	10.3	10.4	1.8
Stage W, %	0.8	1.3	5.9	3.0	3.7	3.8	3.1	1.9

Data are taken from work by J. T. Shurley, C. M. Pierce, and K. Natani (unpublished data on file at the Behavioral Sciences Laboratories, Veterans Administration Hospital, Oklahoma City, Oklahoma 73104, 1969).

TABLE 10. Summary for USARP 1968 South Pole Sleep for Nine Subjects

	TZ1 $N = 8$	SPS $N = 9$	W1 $N = 9$	W2 $N = 9$	W3 $N = 9$	TZ2 $N = 9$	Mean $N = 53$	Standard Deviation
Total record time, minutes	387.4	371.0	396.2	472.3	420.3	420.9	411.3	35.5
Total sleep time, minutes	380.8	364.5	373.2	469.5	400.3	415.0	400.6	38.5
Sleep onset latency, minutes	12.7	6.1	13.4	2.1	17.0	4.6	9.3	5.9
REM onset latency, minutes	95.6	69.2	77.4	73.0	75.6	108.6	83.2	15.4
Number of awakenings	6.1	3.0	5.3	6.8	6.7	5.9	5.6	1.4
Number of stage changes	98.8	123.6	104.8	155.1	95.9	146.2	120.7	25.3
Number of movement arousals	48.0	55.6	48.1	72.4	71.8	47.4	57.2	11.9
Total stage M, minutes	1.9	3.6	3.1	6.1	3.3	3.3	3.6	1.4
Total stage W (0), minutes	4.7	2.8	8.3	7.1	9.6	8.2	6.8	2.6
Total stage REM, minutes	90.6	82.5	90.1	107.8	102.5	100.0	95.6	9.4
Total stage 4, minutes	1.8	4.2	4.4	4.3	2.9	15.7	5.6	5.1
Total stage 3, minutes	15.6	25.2	18.1	27.6	18.1	31.5	22.7	6.3
Total stage 2, minutes	229.7	227.8	225.3	285.8	238.2	229.0	239.3	23.2
Total stage 1, minutes	29.6	17.8	21.8	27.9	24.0	26.6	24.6	4.3
Total stage NREM, minutes	276.6	275.0	269.6	345.7	283.3	302.7	292.0	28.7
Number of REM periods	3.1	3.3	3.7	4.7	3.4	3.7	3.7	0.6
Mean inter-REM onset interval, minutes	111.5	98.6	92.8	93.6	102.4	100.2	99.9	6.8
Mean inter-REM interval minutes	80.3	72.6	69.9	70.1	68.4	74.4	72.6	4.3
Stage REM, %	24.3	22.5	23.9	22.7	25.6	23.3	23.7	1.1
Slow-wave sleep (stages 3 and 4), %	4.4	7.3	5.5	6.1	4.2	12.8	6.7	3.2
Stage 4, %	0.4	0.9	1.0	0.8	0.6	4.5	1.4	1.5
Stage 3, %	4.0	6.3	4.5	5.3	3.6	8.2	5.3	1.7
Stage 2, %	61.5	63.7	60.0	61.9	60.5	54.6	60.4	3.1
Stage 1, %	7.9	4.9	6.1	5.9	6.0	6.5	6.2	1.0
Stage W, %	1.3	0.7	2.0	1.5	2.5	1.9	1.7	0.6

Data are taken from work by J. T. Shurley, C. M. Pierce, and A. T. Joern (unpublished data on file at the Behavioral Sciences Laboratories, Veterans Administration Hospital, Oklahoma City, Oklahoma 73104, 1970).

and the effects of cold air breathing, ventilatory and blood-gas exchange, hematological changes, blood pressure, oral temperature and oxygen consumption, heart rate, and electrocardiograms on these subjects in considerable detail, so that only a limited summary need be given here.

All the men were in good health; the mean age was 31.7 years (range of 19–49 years). Measurements were taken during six periods over 2 years. Base line studies before departure of the volunteers were performed at Quonset Point, Rhode Island, and in Oklahoma City, Oklahoma. Initial measurements at South Pole were taken 3–7 days after arrival, subsequent observations being made after 6–10 weeks, during the tenth month, and during the twelfth month. Final measurements were taken in Oklahoma City 3–6 months after the subjects' departure from Antarctica. In addition to these scheduled tests, spirometry was performed on the volunteers during their antarctic stay in the summer before and after 90 min of outdoor work at -20°F and then again in the winter before and after 5, 15, 30, and 90 min of exposure (snow shoveling) to -95°F. All measurements were made indoors at 65°–74°F.

Results of spirometry showed that forced vital capacity was not altered by prolonged exposure to the antarctic environment. However, forced expiratory volume in the first second, forced expiratory flow rate ($FEF_{25-75\%}$), and maximum voluntary ventilation increased significantly during the year at South Pole and fell to base line values when the men returned to the United States. The $FEF_{25-75\%}$ increased progressively throughout the year at South Pole and then returned to base line. Work at -20°F caused a small decrease in the $FEF_{25-75\%}$, but other indexes were unaltered. No significant changes occurred in any of the determinations after work in the cold at -95°F.

Blood-gas exchange studies revealed that minute ventilation increased initially after subjects arrived at South Pole and continued to rise throughout the year. The physiological dead space initially decreased from 215 to 160 ml and then gradually increased over the year to 230 ml, this increase explaining the increase in minute ventilation as being due to dead space ventilation. Alveolar ventilation, however, increased initially and then remained unchanged. Mean alveolar pO_2 decreased from 108 to 67 mm Hg, this decrease being associated with a decrease in arterial pO_2 from 87 to 53 mm Hg. The mean alveolar-arterial pO_2 gradient decreased significantly from a base line of 19 to 13 mm Hg within a week after the subjects arrived at South Pole and to 8 mm Hg in the sixth to tenth weeks but then returned to base line levels during the tenth month. The mean blood *p*H increased from 7.40 to 7.46 and remained alkalotic throughout the year, whereas the mean arterial pCO_2 decreased from 34 to 27 mm Hg and remained in the hypocapnic range. The mean respiratory rate increased from 13 to 16 and remained at that level. The improved ventilation-perfusion relationships demonstrated here by changes in the dead space and the alveolar-arterial gradients may be significant as an early adaptive mechanism to the acute hypoxia of altitude. The mechanism of the persistent alkalosis was not identified. That acclimatization was associated with a decrease in alveolar ventilation is suggested by the low pCO_2 in the first week, followed by a minimal increase. The changes in the arterial pCO_2 are consistent with previous evidence that ventilatory drive is blunted from hypoxia in elevation-acclimatized natives [*Chiodi,* 1957; *Lahiri et al.*, 1969]. Significant airway obstruction as a result of breathing cold air was not demonstrated in any subject.

Hematocrit levels increased from a mean of 45.5% to 48.6%, and the hemoglobin level increased from 14.9 to 17.5 gm/100 ml, as one would expect for this pressure altitude. Electrophoresis showed no change in hemoglobin type. Mean leukocyte count decreased minimally (and not significantly), but the measurement technique may not have been sufficiently sensitive.

Measurements of carboxyhemoglobin in April (tunnels open) showed ranges of 0% (nonsmokers) to 5.6% (heavy smokers). Repeated measures in August (tunnels closed) and October (tunnels just reopened) revealed values of 4.0–9.0%, so that both smokers and nonsmokers showed increased levels when the tunnels were closed. Oral temperatures declined progressively at South Pole from a mean of 98.3°F (36.8°C) within a week of arrival to 96.6°F (35.8°C) in the tenth month, this decline being associated with a nonsignificant decrease in oxygen consumption. The protein-bound iodine (4.6–5.6 μg/100 μl) demonstrated no change throughout the year. In view of these findings, more detailed metabolic studies would be of great interest, since otherwise normal, spontaneously hypothermic subjects are not known to have been studied previously.

Electrocardiograms revealed right axis deviation of 20° in three of 11 subjects as a gradual change at South Pole. There was an initial rise in systolic blood pressure from 111 to 120 mm Hg in the first week, but no other values differed from base line measures, nor were heart rates, rhythm, or P waves changed. Thus the hypothermia was associated with little hemodynamic effect, but the shift to the right of the QRS axis in three subjects suggests pulmonary hypertension, as previously described at an elevation of 3365 meters (11,400 feet) [*Harris and Hansen,* 1966].

This study has documented our thesis that the major physiological changes are related to high-altitude hypoxia. But persistent alkalosis, hypothermia, and other unexplained metabolic changes may also play a role in altering the performance of personnel who winter over. Despite the introduction of man's own technologically fabricated microclimate, there is yet evidence that some physiological cold acclimatization process does occur in man, and this is of great interest for future investigation.

In another study on the effects of hypobaric hypoxia, *Whitcomb et al.* [1970] measured erythropoietin levels in the plasma and urine of 11 of our subjects who traveled to South Pole before departure and over the first 2 weeks of residence. The patterns of rise and fall of erythropoietin levels in both plasma and urine were consistent with those found in persons who have gone from sea level to an elevation of 3350 meters in other geographic locations.

Joern et al. [1970] reported the results of a study of sleep pattern alterations in two trained subjects very familiar with both the procedure and the locale in an attempt to document the pulmonary physiological changes on ascent to altitude, since they might relate to changes in sleep psychophysiology. A functional relationship between respiration and sleep had been suggested by several investigators [*Magnussen,* 1944; *Sieker et al.*, 1960; *Ingvar and Bülow,* 1963], but no reports could be found in which EEG patterns during sleep were related to blood gas measurements under acute altitude stress. Sleeplessness has been clinically observed regularly as one manifestation of mountain sickness, which disappears after adaptation.

In this study, one subject suffered moderate altitude sickness, his symptoms being insomnia, headache, nausea, anorexia, and shortness of breath, clearing after about 5 days, whereas the other subject had no symptoms or complaints, although he rated his sleep subjectively as 'poor.' Pulmonary response and blood gas response were measured 48 hours after arrival at South Pole and were found to be exactly like those described by *Guenter et al.* [1970]. The arterial pO_2 was found to be 47 and 51 mm Hg, about half the base line values. The symptomatic subject showed a much steeper alveolar-arterial gradient (22.0 mm Hg) than the asymptomatic one (8.0 mm Hg). However, the sleep physiology of both was grossly altered in all three recordings over a 2-week period. Both subjects showed complete loss of both stages of SWS (stages 3 and 4) on all nights, a shortened duration of sleep, a large increase in stage 1 (light) sleep, and a decrease in REM sleep. These findings do not correspond precisely with those in any previously described situation. An explanation, based on the normal physiology of sleep and respiration shown previously by *Ingvar and Bülow* [1963], is offered that elimination of stages 3 and 4 avoids the physiological depression of respiratory exchange associated with those stages, the effect being that the effective hypoxic stress is not further aggravated.

INFECTIOUS DISEASE STUDIES

Since infectious diseases among persons wintering-over at antarctic stations are infrequent, studies relating to infection have largely been ignored. This very paucity of infectious disease processes suggested to J. T. Shurley and H. G. Muchmore that antarctic isolation should be viewed as one end of a continuum, the other end of which is crowded urban living with its associated high level of contagion. The investigators thus conceived a study of white cell populations as an index of adaptation to the relatively sterile polar environment and as a yardstick for comparison with urban temperate zone man.

Blood analyses were performed in the 1967 austral winter on five men at Plateau station, perhaps one of the most isolated spots on the globe [*Muchmore et al.*, 1970]. This eight-man station is centrally located on the Plateau of East Antarctica at an elevation of 3600 meters (11,890 feet). Antecubital venipuncture was performed on the volunteers at Davisville, Rhode Island, before departure, four times during the year at Plateau station, and again at Christchurch, New Zealand, after departure from Antarctica. Except for a few hours' visit by a small traverse party from Vostok (equally isolated near the south geomagnetic pole) on March 3, subjects lived in unbroken social isolation from February 7 to November 18, 1967 (41 weeks). Duplicate cell counts

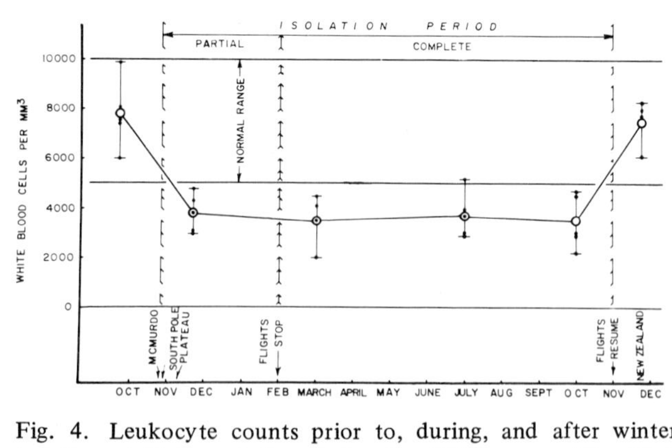

Fig. 4. Leukocyte counts prior to, during, and after winter isolation at Plateau station during the 1967 season. The filled circles are individual determinations, and the open circles are mean values (N = 5). The neutropenia shown is assumed to reflect antigen deprivation secondary to extreme dry cold and unbroken social isolation in healthy men.

and blood smears for differential counts were made (Figure 4).

The total white cell count fell significantly during the period of isolation and returned to base line values within 2 weeks of return to social contact with the outside world. The differential counts showed that the leukopenia originated in a highly significant decrease in the neutrophil series. None of the usual causes of neutropenia were uncovered, nor could the neutropenia be related to the extreme cold, the high elevation, or the prolonged darkness of the 4-month polar night at that location. The investigators concluded that the singularly infection-free condition of wintering-over personnel permitted the granulocyte counts to subside to levels well below those usually regarded as being normal. This agrees with findings reported for animals kept in gnotobiotic (germ-free) laboratory environments for long periods [*Gordon,* 1959; *Miyakawa,* 1959]. The neutropenia was thought to result from a relative deprivation of antigenic material associated with normal social mixing. Thus our current concepts of normal bacterial flora may be only relative, requiring specification of the environmental conditions under which the observations were made. The investigators have suggested that possible altered susceptibility to subsequent infection should be evaluated when isolated humans return to denser population areas.

INDIVIDUAL 'CLINICAL' RESPONSES TO ANTARCTIC RESIDENCE

Since the antarctic environment is inimical to the support of human life and social colonization, some consideration should be given to the adaptation required of the individual who travels or resides there. Many accounts, particularly medical ones, tend to emphasize the dangerous aspects of polar life and to neglect possible healthful and life-sustaining features.

Antarctica is the great heat sink of the globe, rendering its surface incompatible with evolution or maintenance of indigenous biological species. In this respect, conditions for the biosphere differ dramatically from those for the ocean in the tropical zone and are similar to, although they are a good deal less forbidding than, those encountered in outer space and on the surface of the moon and planets other than the earth. Even primitive humans, however, have shown themselves to be uniquely flexible and able to sustain themselves transiently or indefinitely under rather extreme conditions. The technological accomplishments of our species have now permitted a few members to travel in safety to the surface of the moon, to remain there in useful employment for a period, and to return safely to the earth. Conversely, men have gained access to the 'inner space' of the undersea world in submarines or by means of scuba gear, again extending their domain over an alien environment. Similar challenges are presented by Antarctica, the intense cold being chief among these. Man has met them by interposing adequate thermal barriers between his relatively hairless skin and the intense cold air as well as by providing for space heating by means of fossil fuel, electrical, chemical, and atomic energy. In other words, by creating his own portable microclimate, he is able not only to penetrate the antarctic heat sink but also to subsist inside it in relative comfort and safety with mobility. However, the enormously increased cost of 'doing business,' compared with that in long-inhabited parts of the globe, makes a careful cost-reward analysis of the situation essential.

To date, no individual or group has chosen or regarded Antarctica as a possible choice for a permanent home, although a number have clearly chosen to return repeatedly for shorter or longer stays and thereby devoted appreciable segments of their life space to residence under antarctic conditions. Americans call these individuals 'old antarctic explorers,' or 'OAE's'; the Soviets call them 'polar men.' As such, they share many of the experiences of all frontiersmen throughout human history, variations being introduced only by the historical and technological circumstances.

In Antarctica there are no defending or welcoming natives to oppose or encourage one's entry. A multilateral international treaty, the Antarctic Treaty, which temporarily sets aside prior territorial claims and recognizes no new ones for a substantial period of time (perhaps indefinitely), has evolved in the last decade. Prohibition of military operations, conservation of natural resources, freedom of travel, inspection, and information exchange are mutually pledged by the governments of the signatory powers. What is permitted is peaceful pursuit of knowledge by scientists, but tourists are not excluded. So far, no commercially exploitable natural resource has been discovered on the continent, although the surrounding oceans have been stripped of their whale stocks. The two major animal species encountered on the land, seals and penguins, have no fear and impose no threat. No food plant grows there. Except for one nonfatal episode on the Antarctic Peninsula during World War II, no record exists of a single shot fired in anger, nor has any murder or capital crime been committed in Antarctica. History records one individual who was assumed to have taken his own life there. In general then, all threats appear to derive from the natural climatic conditions and the operation of man's technological devices. Accidental injury, primarily from aircraft and land vehicle operation, is by far the chief cause of morbidity and mortality. Diseases have been extremely rare, consisting mainly of those that occur sporadically anywhere (e.g., appendicitis and cardiac and circulatory disorders). Minor dental troubles are not uncommon, and minor mental troubles are both common and temporary. Thus far, resident man has deposited his wastes with impunity to himself and other species except in limited instances of some pollution affecting benthic communities and dry valley microbiota. Cold stress, altitude sickness, hypoxia, and postisolation infectious diseases remain then as the primary physiological threats to man.

In their review of British contributions to polar physiology, *Lewis and Masterton* [1963, p. 1012] stated that

> . . . our studies have now entered a phase where we have perhaps come to the end of using relatively crude methods: we can now offer facilities to specialists who wish to apply their own advanced techniques to the unique opportunities of the Antarctic.

That sophisticated research methods are indeed feasible under the hard, isolated conditions of Antarctica has been demonstrated by the American contributions to polar research reviewed here and reported by *Shurley* [1971].

REFERENCES

Agnew, H. W., W. B. Webb, and R. L. Williams, The effects of stage four sleep deprivation, *Electroencephalogr. Clin. Neurophysiol., 17,* 68–70, 1964.

Aserinsky, E., and N. Kleitman, Regularly occurring periods of eye motility, and concomitant phenomena during sleep, *Science, 118,* 273–274, 1953.

Beazley, R. M., Medical officer at 90° south, *Md. State Med. J., 17,* 52–55, 1968.

Blackburn, A. B., Medical research at Plateau station, *Antarctic J. U. S., 3,* 237–239, 1968.

Brooks, R. E., K. Natani, J. T. Shurley, C. M. Pierce, and A. T. Joern, An antarctic sleep and dream laboratory, in *Polar Human Biology,* edited by O. G. Edholm and E. K. E. Gunderson, pp. 322–341, Heinemann Medical Books, London, 1973.

Chiodi, H., Respiratory adaptations to chronic high altitude hypoxia, *J. Appl. Physiol., 10,* 81–87, 1957.

Committee on Polar Research, *Polar Research: A Survey,* pp. 178–200, National Academy of Sciences, Washington, D. C., 1970.

Dement, W. C., and N. Kleitman, Cyclic variations in EEG during sleep and their relations to eye movements, body motility, and dreaming, *Electroencephalogr. Clin. Neurophysiol., 9,* 673–690, 1957.

Dewan, D. M., The P-hypothesis for dreaming sleep, *Psychophysiology, 5,* 203–208, 1968.

Edholm, O. G., Medical research by the British Antarctic Survey, *Polar Rec., 12,* 575–582, 1965.

Gordon, H. A., Morphological and physiological characterization of germ-free life, *Ann. N. Y. Acad. Sci., 78,* 208–220, 1959.

Gottschalk, L. A., and G. C. Gleser, *The Measurement of Psychological States through Content Analysis of Verbal Behavior,* pp. 1–317, University of California Press, Berkeley, 1969.

Guenter, C. A., A. T. Joern, J. T. Shurley, and C. M. Pierce, Cardiorespiratory and metabolic effects in men on the South Polar Plateau, *Arch. Intern. Med., 125,* 630–637, 1970.

Gunderson, E. K. E., Emotional symptoms in extremely isolated groups, *Arch. Gen. Psychiat., 9,* 362–368, 1963.

Gunderson, E. K. E., Mental health problems in Antarctica, *Arch. Environ. Health, 17,* 558–564, 1968.

Hall, C. S., and R. L. Van De Castle, *The Content Analysis of Dreams,* pp. 1–320, Appleton-Century-Crofts, New York, 1966.

Harris, C. W., and J. E. Hansen, Electrocardiographic changes during exposure to high altitude, *Amer. J. Cardiol., 18,* 183–190, 1966.

Hunt, R. B., Clinical observations on adaptation to antarctic life, *Mil. Med., 133,* 625–628, 1968.

Ingvar, D. H., and K. B. Bülow, Respiratory regulation in sleep, *Ann. N. Y. Acad. Sci., 109,* 411–448, 1963.

Joern, A. T., J. T. Shurley, R. E. Brooks, C. A. Guenter, and C. M. Pierce, Short-term changes in sleep patterns on arrival at the South Polar Plateau, *Arch. Intern. Med., 125,* 649–654, 1970.

Kleitman, N., and H. Kleitman, The sleep-wakefulness pattern in the Arctic, *Sci. Mon., 76,* 349–356, 1953.

Lahiri, S., N. H. Edelman, N. S. Cherniack, and A. P. Fishman, Blunted hypoxic drive to ventilation in subjects with life-long hypoxemia, *Fed. Proc. Fed. Amer. Soc. Exp. Biol., 28,* 1289–1295, 1969.

Lewis, H. E., Sleep patterns on polar expeditions, in *The Nature of Sleep, Ciba Foundation Symposium,* edited by G. E. W. Wolstenholme and M. O'Connor, pp. 322–328, Little, Brown, Boston, Mass., 1961.

Lewis, H. E., and J. P. Masterton, Sleep and wakefulness in the Arctic, *Lancet, 1,* 1262–1266, 1957.

Lewis, H. E., and J. P. Masterton, Polar physiology: Its development in Britain, *Lancet, 1,* 1009–1014, 1963.

Magnussen, G., *Studies on Respiration during Sleep: A contribution to the Physiology of Sleep Function,* pp. 1–276, H. K. Lewis, London, 1944.

Miyakawa, M., The lymphatic system of germ-free guinea pigs, *Ann. N. Y. Acad. Sci., 78,* 221–236, 1959.

Muchmore, H. G., A. B. Blackburn, J. T. Shurley, C. M. Pierce, and B. A. McKown, Neutropenia in healthy men at the South Polar Plateau, *Arch. Intern. Med., 125,* 646–648, 1970.

Mullin, C. S., Jr., Some psychological aspects of isolated antarctic living, *Amer. J. Psychiat., 117,* 323–327, 1960.

Natani, K., and J. T. Shurley, Sociopsychological aspects of a winter vigil at South Pole station, in *Human Adaptability to Antarctic Conditions, Antarctic Res. Ser.,* vol. 2, edited by E. K. E. Gunderson, this volume, AGU, Washington, D. C., 1974.

Natani, K., J. T. Shurley, C. M. Pierce, and R. E. Brooks, Long-term changes in sleep patterns in men on the South Polar Plateau, *Arch. Intern. Med., 125,* 655–659, 1970.

Newman, E. A., and G. R. Evans, Human dream processes as analogous to computer program clearance, *Nature, 206,* 534–536, 1965.

Pratt, H. S., E. P. Beck, L. S. Wirthlin, and A. Graybiel, Studies on the response to acute altitude exposure with special reference to the possibility of early detection of high altitude pulmonary edema, *Rep. NAMI-964,* U.S. Navy Aerosp. Med. Inst., Pensacola, Fla., May 1966.

Priestley, R., Inexpressible Island, *Nutr. Today, 4,* 18–27, 1969.

Rechtschaffen, A., and A. Kales (Eds.), A manual of standardized terminology, techniques, and scoring system for sleep stages of human subjects, *Publ. 204,* 57 pp., Public Health Serv., Nat. Inst. of Health, Bethesda, Md., 1968.

Sassin, J. F., D. C. Parker, J. W. Mace, R. W. Gotlin, L. C. Johnson, and L. G. Rossman, Human growth hormone release–Relation to slow-wave sleep and sleep-waking cycles, *Science, 165,* 513–515, 1969.

Selye, H., *Stress,* Acta Inc., Medical Publishers, Montreal, 1950.

Serafetinides, E. A., J. T. Shurley, and R. E. Brooks, EEG and the effects of photic stimulation at the south pole, *Aerosp. Med., 42,* 52–53, 1971.

Shurley, J. T., Dreams good and bad, *Nutr. Today, 5,* 30–31, 1970*a*.

Shurley, J. T., Man on the South Polar Plateau, *Arch. Intern. Med., 125,* 625–629, 1970*b*.

Shurley, J. T., New directions in antarctic biomedical research, in *On Research in the Antarctic,* edited by L. O. Quam, pp. 277–287, AAAS, Washington, D. C., 1971.

Shurley, J. T., C. M. Pierce, K. Natani, and R. E. Brooks, Sleep and activity patterns at South Pole station, *Arch. Gen. Psychiat., 22,* 385–389, 1970.

Sieker, H. O., A. Heyman, and R. I. Birchfield, The effects of natural sleep and hypersomnolent states on respiratory function, *Ann. Intern. Med., 52,* 500–516, 1960.

Sladen, W. J. L., Staphylococci in noses and streptococci in throats of isolated and semi-isolated antarctic communities, *J. Hyg., 63,* 105–116, 1965.

Taylor, I. M., Medical experiences at McMurdo Sound, in *Cold Injury, Transactions of the Sixth Conference,* edited by S. M. Horvath, pp. 117–140, Macy, New York, 1960.

Tyler, P. E., Medical aspects of the U.S. antarctic program, *Antarctic J. U. S., 3,* 233–237, 1968.

Webb, W. B., Measurement and characteristics of nocturnal sleep, in *Progress in Clinical Psychology,* vol. 8, edited by L. E. Abt and B. Reiss, pp. 2–27, Grune and Stratton, New York, 1969.

Whitcomb, W. H., A. T. Joern, C. A. Guenter, M. Moore, J. T. Shurley, and C. M. Pierce, Effect of the South Polar Plateau on plasma and urine erythropoietin levels, *Arch. Intern. Med., 125,* 638–645, 1970.

Wilson, O., Human adaptation to life in Antarctica, in *Biogeography and Ecology in Antarctica, Monogr. Biol.,* vol. 15, edited by J. van Mieghem, P. van Oye, and J. Schell, pp. 690–752, W. Junk, The Hague, Netherlands, 1965.

SOCIOPSYCHOLOGICAL ASPECTS OF A WINTER VIGIL AT SOUTH POLE STATION

KIRMACH NATANI

Behavioral Sciences Laboratories, Veterans Administration Hospital, Oklahoma City, Oklahoma 73104

JAY T. SHURLEY

Behavioral Sciences Laboratories, Veterans Administration Hospital, Oklahoma City, Oklahoma 73104
Department of Psychiatry and Behavioral Sciences, University of Oklahoma Health Sciences Center
Oklahoma City, Oklahoma 73190

Abstract. Human responses to life in the closed and isolated environment that exists for 8 months annually at the U.S. Amundsen-Scott South Pole station are presented, compared, and evaluated within the context of a review of previous studies at both South Pole station and other station sites in Antarctica. The investigators on one hand have the integrative perspective of those who are intimately familiar with the literature and on the other have actually experienced and observed the conditions in the field while they were conducting their studies. The findings both confirm and amplify previous reports of human behavior in a demanding environment characterized by extreme cold, markedly different light-dark cycles, hypobaric hypoxia, lack of novelty, and the social stresses encountered by a collection of heterogeneous strangers in developing a distinct microculture adapted to this unusual human situation. It is the unusual isolation of the group from the normal constraints of society and the simplicity of the physical conditions that suggest that the insights gained in this natural laboratory may have direct application to the problems of individuals isolated, stigmatized, or culturally distinct within the mundane routine of modern industrialized society. The authors interpret the negative aspects of life in Antarctica as the normal responses of healthy men to unusual and extreme stresses. Both the conditions and the responses remind us that here the major role and responsibility of the behavioral scientist are to assure the selection of healthy men for such endeavors and at the same time to remain alert to the existence and possible development of physical and social conditions that could produce pathological responses. They choose to emphasize the often neglected problem of measuring positive responses and assessing human well-being in the face of adversity. They have attempted to study responses to the social and psychological environment as well as physiological responses to the unique geophysical conditions and to define their total influence on human health on the South Polar Plateau. The preliminary findings summarized here suggest that the combined methodologies of the social psychologist and the psychophysiologist can be very productive in this type of field study. At the same time, they indicate that future investigations should employ the more global and integrative approach of the medical biometeorologist and should concentrate on psychophysiological and neuropsychological measures of responses to both the social and the physical aspects of the environment.

Antarctic stations provide an unusual opportunity to study a small, closed group of human beings in an isolated, stressful environment. Of course, many situations exist in which individuals are separated from their accustomed surroundings, e.g., students in college dormitories, missionaries in remote regions, and men in military service. But the 'input' from the new environment and the social processes under such circumstances are too varied to study the effects of separate factors on the individual. True, these variables still cannot be controlled in the antarctic station, but the complexity of their interactions is greatly diminished, and their boundaries are better defined. Here station crews live in complete, unbroken geographic isolation during the antarctic winter, when neither departure nor physical access is possible at the inland stations for 8- to 10-month stretches of time owing to the extreme cold of the antarctic winter. The mean midwinter temperature of the South Polar Plateau, for example, is -62°C (-79.6°F). At present levels of technologic development, both surface travel and air travel are virtually impossible under such conditions.

This southern outpost has been staffed over the winter largely by volunteer crews, consisting of 16–24 men. Although radio communication with other antarctic stations, the United States, and other parts of the world is usually well maintained, there are times, during periods of intensified solar activity, when even radio communication is impossible because the ionosphere will not reflect the signals back to earth.

Despite the obvious potential for field research into the effects of isolation on small groups, few studies of this sort have been reported in the growing body of small-group literature. Thus the potential for social science research in the Antarctic remains relatively untapped. Perhaps this lack of research is due, as *McGrath and Altman* [1966] indicate, to the traditional academic orientation of small-group researchers and their reliance on laboratory control, a relatively small number of variables, the availability of cooperative college subjects, and pressure to complete the project rapidly. One must consider still another factor, however: the hurdles that must be overcome in obtaining psychological data in the Antarctic. Traditionally, psychologists have been viewed with suspicion and have been offered minimal cooperation by antarctic volunteers. *Siple* [1959] first commented on the disruptive effect of sociometric tests on station morale. *Nardini et al.* [1962, p. 100] reemphasized this problem, stating that, 'The peer nomination technique generally proved to be so psychologically threatening that it was discontinued as a routine.' More recently, *Taylor* [1969*b*] described the disturbing effect that a psychologist's presence may have for these small groups in any role that he may assume.

THE VOLUNTEER: INCENTIVES FOR ANTARCTIC DUTY

The prospect of scientific discovery is the raison d'être of the antarctic stations [*Gunderson,* 1968]. No commercial or economic activity of any significance has yet been established on this continent. High on the list of reasons that individual men give for volunteering are earning and saving money [*Nardini et al.,* 1962; *Rohrer,* 1961; *Taylor,* 1969*b*]. *Law* [1960] observed that financial anxiety is absent at antarctic stations because money is not used. Other common motivations mentioned by the men are broadened experience [*Law,* 1960], advancement [*McGuire and Tolchin,* 1961; *Nardini et al.,* 1962], prestige, intellectual curiosity [*Rohrer,* 1961], increased technical knowledge, and independent action [*Taylor,* 1969*b*]. Motivations often unstated but nevertheless common are the desire for adventure, challenge, gratification of ambition, the opportunity to view magnificent scenery, and the need to prove or find oneself [*Law,* 1960; *Nardini et al.,* 1962; *Rohrer,* 1961; *Taylor,* 1969*b*]. The antarctic experience may be part of a continuing search for self-identity, or it may reactivate this process in the individual.

At the same time, these investigators have identified escape as an underlying motivating factor. *Law* [1960] has examined this feature of antarctic life in detail. He pointed out that everyone's status depends primarily on his work efficiency; hence there is no concern over preconceived ideas of social status. Social graces cease to have significance, the antarctic station thus becoming a haven for the technically competent individual who is deficient in social skills. *Taylor* [1969*a*] supports this observation with his own findings that many New Zealand volunteers are 'professional isolates' who regularly move from one isolated duty station to another. *Nardini et al.* [1962] found that some servicemen volunteer to escape marital conflicts and what they consider to be less desirable duty stations. *Law* [1960] emphasized the attractiveness of the lack of social distractions at an antarctic station. There one has no obligations to family and friends and not as many mandatory social roles to play. Time is not consumed by commuter travel, weekends, or holidays, and thus it is possible to accomplish 2 normal years of work in 1.

On the other hand, those men who choose to return to the Antarctic may also seek the special qualities of family living to be found there when effective, stable, coherent, well-functioning groups develop. In the Antarctic, bonds of mutual concern and obligation as well as those of common interests and values allow the whole sum of experience to be tied together and shared. These processes of intimacy have been seriously restricted and enfeebled by modern metropolitan society (B. De Voto, introduction for *Homans* [1950, p. xvi]), and their presence in the society of man in Antarctica may provide an incentive for those who return or become the professional isolates. The favorable responses of men to a sojourn in an isolated or high-risk environment are discussed in the section on positive aspects of antarctic wintering.

PHYSICAL ENVIRONMENT

Gunderson and Nelson [1963] commented on the monotony of station routine and of stimuli in general. As *Taylor* [1969*b*] pointed out, a sense of paucity, even sterility, pervades the polar station: there are no green growing plants to see or smell, no fresh foods to taste, no new clothes to wear, and no sounds of nature to hear (the sound of falling rain, the rustling of leaves, etc.). Although the antarctic continent covers 14.24 million km^2 (5.5 million mi^2), only 10,350 km^2 (4000 mi^2) is estimated to be suitable for sustaining life. The largest permanent inhabitant of the continent is a tiny fly. Essentially no life can exist inland beyond the narrow coastal regions without artificial support. There are no land vertebrates or birds, no amphibians, no reptiles, no freshwater fish, no mollusks, and no earthworms. Vegetation is virtually absent except for a few simple life forms such as lichens, mosses, and fungi in protected ice-free coastal areas [*Francis and Mason,* 1965].

Moreover, the South Polar Plateau suffers from a unique constellation of adverse climatological factors: high elevation, extreme cold, extreme dryness, and the absence of accustomed diurnal light-dark stimuli. At the south pole, for example, the elevation is only a little over 2800 meters (9000 feet) above sea level, but because of the thinning of the earth's atmospheric shell at the poles the atmospheric pressure is equivalent to an elevation of approximately 3353 meters (11,000 feet) in temperate zones. The mean annual temperature is -51°C (-59.8°F), and temperatures range from -17°C (+1.4°F) down to -85°C (-121°F). The relative humidity is too low to permit accurate measurement. Finally, the annual seasons consist of one continuous day (summer) and one continuous night (winter), each lasting approximately 6 months.

Not only is the external environment hostile, but the life within a small antarctic station itself is characterized by lack of accustomed comforts, few luxuries, and a monotonous diet. Most of the original International Geophysical Year stations have largely been rebuilt with modern materials and construction techniques that greatly ameliorate the effects of the climate and provide better facilities in which to work and live. A new South Pole station structure is now scheduled for occupancy in 1975. However, the descriptions of the original stations [*Francis and Mason,* 1965, p. 391] are basically still valid for South Pole station:

> . . . they were crude affairs which leaked and sagged as the snow melted from above and below. Their ventilation was bad, their storage areas inadequate and their electrical systems dangerous and undependable. The scientific working space was crowded, makeshift and badly lit and heated.

Recreational facilities also are generally inadequate at the small stations [*Gunderson,* 1966*b, c*]. Still, undue emphasis should not be placed on crude facilities and harsh physical conditions in the environment. *Haythorn and Altman* [1966] indicated that sociopsychological hazards may equal physical threats in extreme environments. *Lantis* [1968] went as far as to implicate specifically the social environment as the most potent source of stress in the polar regions. A recent report by *Gunderson* [1968] supports her contentions. Gunderson found that improved living and working conditions have not significantly reduced the incidence of physical complaints and symptoms at antarctic stations. Social stresses and the symptoms most often reported, characteristically of a psychosomatic nature, are discussed in the section on negative aspects of antarctic wintering.

PSYCHOLOGICAL ENVIRONMENT

Law [1960, 1963] reported that a marked feature of the psychology of antarctic station personnel is the hunger for recognition and praise exhibited by many of the men. So sensitive do the men become that even shortages in station supplies or a delay in the arrival of the relief party may be interpreted as a sign of rejection by outside superiors. The result is a drop in individual and group morale. *McGuire and Tolchin* [1961] commented that the men could easily become demoralized by the imposition of unfeeling outside authority on them and the feeling that there was a lack of adequate communication between their station and the United States as well as between their station and other antarctic stations.

Gunderson [1968] approached the problem of rewards, sense of personal significance, and subjective feelings of usefulness in both civilian and Navy personnel stationed in Antarctica. He indicated that the civilian scientists expected to receive delayed professional rewards and satisfactions from their work. The Navy men, on the other hand, were dependent on more immediate rewards, such as the approval of their associates and supervisors, promotion in rank, and a desirable assignment for their next duty station.

The psychological deprivations threatening station personnel involve temporary reductions in the gratification of the basic human needs of affection, security, and feelings of personal significance. *Rohrer* [1961] emphasized status leveling (due to loss of conventional status symbols) and the loss of social roles, such as those of husband, brother, father, and club member, that are available in normal society. *Wilson* [1965] noted that sexual deprivation is often stressed in antarctic reports. *McGuire and Tolchin* [1961] stressed the actual physical threats leading to psychological insecurity, such as the constant dangers of fire, frostbite, and getting lost in the dark. These threats seemed more ominous for their group because most felt that their survival training had been inadequate and was previously untested.

Rohrer [1961] believed that three major threats face the man stationed in Antarctica: harm from the physical environment, social isolation from the station party, and personal psychic harm resulting from the isolation experience.

NEGATIVE ASPECTS OF ANTARCTIC WINTERING

Antarctic psychological research has shown that men who 'winter over' become increasingly susceptible to insomnia, disrupted sleep patterns, anxiety, depression, and irritability. *Nelson* [1963] has reported that individual differences are sufficient to warrant using these symptoms as indexes of emotional adaptation. Both *Gunderson* [1963] and *Nelson* [1963] have suggested that such symptoms, indicative of heightened levels of anxiety and problems of emotional control, are likely to sap a man's motivation and hinder his social adjustment. This is particularly true at the small antarctic stations, where men find it extremely difficult to conceal their feelings from their companions.

Soviet reports have reversed the cause-effect relationship by suggesting that it is social factors that determine sleep patterns. V. N. Semagin (cited by *Wilson* [1965]) considers social factors to have a significant influence on sleep patterns in the Arctic. *Bogoslovskii* [1973] has made a similar conclusion for Soviet coastal stations in Antarctica and reports that he found no 'polar insomnia' when he was using actographic methods to study sleep.

The antarctic station is a total institution. Total institutions are defined by *Goffman* [1961] as any place where a group of individuals suffering a common fate resides and works while it is separated from society for an extended period of time. A feature common to all closed institutions is that they change people, either by design or by accident. The total institutions of normal society represent a formalized way in which various stresses are applied to individuals to effect planned changes in their characteristic modes of response to their environment. These stresses may be both physical and social. But social stresses appear to be the most important agents of change; they include: devaluation of the self, social deprivation, enforced socialization, status incongruency, sociocultural complexity, inadequate opportunity for personal expression, development of identity conflicts, cognitive dissonance, and constant threat of social rejection [*Lantis,* 1968]. At an antarctic station all the men are subjected to some of these stresses at one time or another. Some men may struggle with them for the entire year, and social pressures such as these, as well as the absence of interesting stimulus variations in the environment, may be primarily responsible for the symptoms generally reported from all the stations.

The symptoms listed below have been found to some extent in every group of men wintering in Antarctica. They are: (1) insomnia, popularly known as 'big eye'; (2) irritability; (3) headache; (4) nightmares; (5) anxiety; (6) mild depression; (7) boredom; (8) fatigue; (9) decline in personal hygiene; (10) reduced motivation combined with intellectual inertia, impaired memory, impaired concentration, decline in alertness, and a general apathetic state; (11) increased appetite, frequently accompanied by weight gains; (12) digestive ailments; (13) rheumatic aches and pains; and (14) increased sensitivity to physical and social stimuli [*Gunderson,* 1963; *Il'in and Egorov,* 1970; *Il'in,* 1973; *Law,* 1960; *McGuire and To'chin,* 1961; *Mullin,* 1960; *Nelson,* 1963; *Nardini et al.,* 1962; *Palmai,* 1963*a, b*; *Rohrer,* 1961; *Taylor,* 1969*b*]. Many of the same symptoms have been reported in men wintering in the Arctic [*Boag,* 1952; *Eilbert and Glaser,* 1959] and in men confined in a concentration camp [*Kral,* 1951]; similar symptoms have been noted during mourning [*Haggard,* 1964]. The extreme sensitivity of the men to any apparent rejection by outside authorities, such as a late relief party [*Law,* 1960], appears to be highly related to the emotional sensitivity present in mourning. *Janis* [1969, p. 189] said:

> In part this sensitivity seems to stem from the arousal of deep-seated dependency needs and involves a reactivation of childhood fears of being abandoned . . . . Thus, any . . . apparent form of neglect, is likely to evoke an emotional reaction out of all proportion to the circumstances.

Rohrer [1961] noted a phenomenon peculiar to men whom he thought had been socially rejected for one reason or another by the station party. He labeled it the 'long eye,' described elsewhere as 'a twelve-foot stare in a ten-foot room' [*Haggard,* 1964, p. 457]. *Rohrer* [1961] explained the long-eye syndrome (i.e., sleeplessness, hallucinations, breaking down in tears, loss of appetite, prolonged silence, suspiciousness, reduced work, etc.) in terms of sensory deprivation. Laboratory sensory deprivation studies of groups in confined isolation, reviewed by *Wheaton* [1959], *Ruff* [1961], and *Ruff et al.* [1959], have revealed the emergence of similar symptoms. However, the long-eye stare phenomenon, reported in isolation context, remains peculiar to the Antarctic. Personal observations by K. Natani in Antarctica led him to hypothesize that the long-eye phenomenon may bear a close relationship to the staring reaction found in individuals who have witnessed large-scale disaster [*Janis,* 1969]. Speaking of disaster witnesses, *Janis* [1969, p. 191] noted:

> . . . their mood resembled that of grief-stricken mourners. Their thoughts were directed toward trying to comprehend the significance of the disaster, not in terms of its physical causes, which had already been explained to them, but in relation to their personal concepts of life and death.

The long eye has recently been dichotomized into two discrete behavioral phenomena, 'staring' and 'drifting,' by *Popkin et al.* [1974], who presented the first formal descriptions of these two altered states of consciousness. They reported that staring is a transient and episodic phenomenon that appears to be associated with leisure activity in the evening, whereas drifting may appear as an ongoing process with concomitant alterations in memory, speech, and emotional control. Both behaviors seem to be limited to the latter part of the antarctic year.

POSITIVE ASPECTS OF ANTARCTIC WINTERING

Most antarctic reports have tended to cast antarctic living in a primarily negative light. Yet this view is inconsistent with reactions reported by volunteers in the field. *Mullin* [1960] interviewed men who had wintered over and reported that the majority felt that something good had happened to them during their year of isolation. They expressed feelings of having developed greater self-discipline, adaptability, tolerance, patience, and self-understanding.

In their field research on individuals isolated in an undersea habitat (i.e., SEALAB), *Radloff and Helmreich* [1968, p.98] made the following observation:

> . . . it is more often the case that volunteers who undergo the high stresses of real-life situations tend to react more favorably than do men experiencing the lesser objective stresses of laboratory and simulation studies. Men and crews frequently maintain high levels of performance, establish close and warm interpersonal relations, and remember the experience as an outstanding positive event in their lives. Furthermore, there appears to be very little, if any, correlation between severity of stress and favorability of reactions within field studies.

Radloff and Helmreich explained these discrepancies between negative regard and positive regard for experiences with a cost-reward analysis of the total social system in which individual reactions occur. The trouble with this technique is that positive as well as negative factors must be measured and scaled before the outcome for an individual may be deduced. The overabundance of negative factors reported in most field data suggests that the measures presently used are inadequate or inappropriate.

Negative emotions are relatively easy to identify, whereas positive feelings, tending to be based on personal evaluations of success and failure, are somewhat nebulous. At the moment, how to make an adequate assessment of positive responses remains an unsolved problem. It seems easier to ask people about their troubles than to tap the positive satisfactions that they are experiencing. *Bradburn and Caplovitz* [1965] in field research unrelated to the antarctic studies indicated that positive and negative feelings lie on independent dimensions and that knowing an individual's response level for negative feelings does not allow one to predict his level for positive feelings. They studied men in three small midwestern communities to determine the effects of differing degrees of environmental stress on what has variously been called mental health, subjective adjustment, happiness, or psychological well-being. They focused on anxiety, marital adjustment, job adjustment, and social interactions in these men (ages 25–49) over an 8-month period.

Wheaton [1959] believes that negative reactions to isolation per se are reduced within a well-functioning group and that deviations from normal behavior can effectively be prevented with proper psychological preparation, training, self-discipline, and definite regulation of daily activities. *Wilkins* [1952, p. 449] lends support to this with the statement that, 'The problems of the white man in the Arctic are not all psychiatric or psychological, but rather [include] problems of indoctrination, training, and administration.'

Finally, *Wilson* [1965] cautioned against the overemphasis of psychological problems and suggested that future research concentrate on determining what proportion of the men wintering in Antarctica adjust well to the environment and why.

Taylor and Shurley [1971] used an objective personality test (R. B. Cattell's 16 personality factor test) to verify the positive subjective reports obtained by *Mullin* [1960] in postisolation interviews. Their results indicate that, after wintering at South Pole, men are significantly more self-sufficient, more self-controlled, and calmer and have a tendency to be silent but open to fresh ideas. Perhaps they could be described as more mature in the sense of being autonomous, i.e., more independent of environmental influences. They tend to be influenced less by social pressures and are also characterized by their exercise of due care and consideration in their social behavior.

Individuals who have spent some time in the Antarctic like to refer to themselves, especially when they are with men of a similar disposition, as 'old antarctic explorers' (OAE's). This facet of the Antarctican's behavior suggests that a positive sentiment exists in this group with regard to a common social fate such as one finds among the graduates of private schools and some graduate colleges. *Taylor's* [1969*a*] comments on New Zealand volunteers who have returned to the Antarctic on one or more occasions appear to be generalizable to the men of other nations at stations in Antarctica. He has reported that the activities of these men have established the reputation of a corps d' elite for their group, and they provide the mainstay for various antarctic wintering parties as well as for summer field operations.

If in general we can assume that the OAE is an individual who has adapted successfully to life in Antarctica, then we can hypothesize that the OAE has probably learned to form a complex impression of other individuals that includes tolerance for more deviant behavior in others as well as himself. This may make it possible for him to interact with other people more effectively and as a consequence may make his social encounters and relationships more enjoyable, effective, and creative.

THEORETICAL ANALYSES OF HUMAN BEHAVIOR UNDER ISOLATED CONDITIONS INCORPORATING ANTARCTIC DATA

Few theoretical articles have been published on the unique circumstances that men encounter in spending a winter in Antarctica. Apparently only two authors, D. A. Strickland and E. A. Haggard, have attempted theoretical analyses of behavior in isolation incorporating antarctic data.

Strickland [1964, 1965], a political scientist, reviewed certain aspects of human behavior under conditions of threat and isolation, particular emphasis being placed on the antarctic work of E. K. E. Gunderson and P. D. Nelson. He then analyzed group behavior with respect to questions of substantive and procedural authority, the generation of conflicts (e.g., between mission-oriented civilian scientists and support-oriented military personnel), and conflict resolution in a closed environment where there are no places for dissidents to go, no alternative groups to join, and perhaps no external authorities to appeal to in disputes with the local leader.

He noted that some of the possible consequences of such a situation are insubordination, mutiny, inefficiency, low morale, self-defeating and suicidal acts, incapacitation through illness, inflexible and maladaptive behaviors, perceptual distortions, sabotage, and serious logistic error [*Strickland,* 1964]. Small-group researchers, he suggested, have neglected the dynamic and functional aspects of small-group interactions [*Strickland,* 1965]. He further contended that social scientists have generally paid little attention to the empirical study of conflict within small groups, and only rarely have the politics of intermediate-sized groups in a closed environment been reported in the literature.

After examining the data of Gunderson and Nelson, Strickland indicated that a major source of conflict at U.S. antarctic stations is the inherent difference between the disciplined, regulated, conservative activities of the Navy personnel on the one hand and the relatively unstructured, disorderly, independent life-styles of the scientists on the other.

In the same vein, *Mullin* [1960] indicated that hostility was vented by Navy personnel in horseplay, swearing, complaining, and exchanging insults without arousing angry reactions. Among the scientists, however, who are less inclined to use horseplay and verbal exchanges in dissipating their hostility, he implicated repressed hostility as a possible source of headache.

Haggard [1964], the second theoretician, prepared an extensive review of isolation in a myriad of physical contexts. These included individual and group experimental isolation and voluntary and involuntary field isolation for both individuals and groups. He examined the psychological effects of isolation in terms of personality stability and environmental factors that would be expected to produce personality changes. His conclusion was that man usually cannot adapt to extreme or extended isolation without showing some of the symptoms that typically characterize the mentally ill. He suggested that it is the absence of familiar, meaningful objects and relationships, as well as the presence of the unfamiliar, that produces the disturbances associated with isolation. Haggard postulated that adaptation is facilitated if the individual can engage in activities that structure his time and thoughts and also contribute to his well-being. Moreover, after successful adaptation, basic positive changes in personality and mental life are likely.

ISOLATION AND ACTIVITY: THE PROBLEM OF LEISURE TIME

The role of activities in structuring time and thoughts and in facilitating adaptation to a new environment, as described by Haggard, received little attention until the advent of manned space flight and the eventual prospect of long-duration space missions [*Eddowes,* 1961; *Eberhard,* 1967; *Fraser,* 1968].

Harris [1967, p. 115] presented an explicit statement of the problem of time use for the average individual in a normal environment:

> For many people the pressing question is, 'How am I going to get through the next hour?' The more structured time is, the less difficult is this problem. Very busy people with many external demands do not have time on their hands. The 'next hour' is very well programed. This programing, or structuring, is what people try to achieve, and when they are unable to do it themselves, they look to others to structure time for them. 'Tell me what to do.' 'What shall I do next?' 'What we need is leadership.'

Shneidman [1967] divided human behavior into three basic categories depending on how an individual uses time. He postulated that a given time period (i.e., days, months, and years) may be evaluated in terms of the quality of its use and the way in which it has influenced the ongoing life of the individual. His time use categories consist of: (1) elevative patterns of behavior, which tend to increase individual happiness, promote greater maturity, stimulate better interpersonal relationships, and lead the individual toward self-actualizing activities (Shneidman commented that elevative moments are rare in most lives); (2) temporative patterns of behavior, characterized primarily as ways of passing time when the quality of the individual's live is neither elevated nor degraded (e.g., spectator sports, reading the newspaper, and social rituals); and (3) reductive or inimical patterns of behavior, which degrade the quality of life or truncate the length of an individual's life, such as patterns of repeated failure in goal achievement, alcoholism, and drug addiction.

Although Shneidman did not specifically relate time use and mood, he implied that reductive behavior is the result of a maladaptive response to environmental crisis. His observations suggest that patterns of time use and social interaction might provide a means for identifying individuals who are better adapted to antarctic station life. Life, after all, is how one organizes his experiences in time, and behavioral scientists need to relate the manner in which men spend their time individually to their success as group participants [*Pierce,* 1971].

Two adaptive behaviors that may be observed in Antarctica have been described by Haggard and Palmai. *Haggard* [1964, p. 454] stated: 'By and large, emotionally healthy and mature individuals with active, imaginative minds who can become absorbed in such pastimes as reading and who can structure their thoughts and activities adapt best to isolation.' *Palmai* [1963*a*] found that 'social isolates' were better adjusted, were more tolerant of others, used their leisure time more profitably, and made significantly fewer sick calls. Palmai defined as social isolates those men who do not become closely associated with groups larger than a dyad.

LEISURE TIME USE IN ANTARCTICA

Previous antarctic reports have largely bypassed the problem of off-duty time use, although relationships have been noted between off-duty activities and performance, social compatibility, and group size. Yet the most difficult task for many men in the Antarctic appears to be filling their time in meaningful ways [*Taylor and Shurley,* 1971].

Apparently, only one report has been published that specifically deals with the use of leisure time at antarctic stations. *Doll and Gunderson* [1969*b*] investigated hobby interests and leisure activities reportedly liked or disliked by three disparate occupational subgroups and the consistency of these interests. Their results are discussed in detail by *Gunderson* [1974].

Recently, *Natani* [1971] explored possible relationships between the way in which a man spends his leisure time at a small antarctic station and his mood and performance. He correlated nightly observations of individual leisure time activities with a rated monthly level of adaptation to antarctic station life. In rating mood, he defined positive or negative affect as pleasant or unpleasant feelings about being in Antarctica. *Shneidman's* [1967] time use categories were used to evaluate performance, and the individual's time use choices were recorded. It was assumed that temporative leisure time choices indicate an average performance level, whereas reductive time use reflects impaired performance. Adaptation level was rated in terms of this net affect: the ratings were positive, negative, or neutral as determined by specific behaviors.

Natani studied the same leisure activities evaluated by Doll and Gunderson in addition to other time use behaviors described in previous antarctic reports as being indicative of positive adaptation. The observations were made between 10 P.M. and 8 A.M. every day, at times when one would normally expect the men to be sleeping. It was assumed that what a man does during unstructured time when he would normally be sleeping may be highly related to his psychological state at the time. Observations taken during the day by a participant observer may be insensitive to adaptation level because this time is usually structured at the sociocultural level, e.g., by work and meals. The results of this study are described, beginning with the section on participant observation at 90°S.

PARTICIPANT OBSERVERS IN ANTARCTICA

Before *Shurley* [1970, 1971] initiated the present trend in antarctic research of sophisticated global field investigations of sleep psychophysiology and activity cycles, participant observers had published five reports of their experiences and observations in the isolated antarctic environment. Two of these behavioral studies were performed during the antarctic summer (i.e., late October, November, December, January, and early February) by *Smith* [1966] and *Shurley et al.* [1970]. Smith observed subgroup formation and structure in a seven-man traverse party during periods of danger and boredom on the Polar Plateau. Shurley et al. used daily self-reports from 32 men at South Pole station to describe how time was used for a typical week of normal summer operation at the station. The remaining three studies took place during the antarctic winter at the U.S. South Pole station [*McGuire and Tolchin,* 1961], an Australian station at Macquarie Island [*Palmai,* 1963*a, b*], and the Japanese Shōwa station [*Matsuda,* 1964].

Before the recent work of Shurley and co-workers, *Palmai* [1963*a, b*] apparently performed the most methodologically thorough participant observation study of behavior in Antarctica. Palmai's data on 14 men consisted of 1-hour tape recordings of group discussions taken twice each week between 7 P.M. and 8 P.M., incidental records of interpersonal behavior, a quarterly interpersonal checklist, quarterly records of obvious cliques and mutual preferences, and personal reactions and symptoms reported to the medical officer. The data were analyzed in terms of: (1) emotional interactions between the participants during discussions (R. F. Bales' interaction process analysis), (2) the 'meaning' of 30 descriptive terms used in self-description or description by others for each man for each quarter of the winter (C. E. Osgood's semantic differential), (3) the behavior traits exhibited during each quarter (T. Leary's interpersonal checklist, e.g., managerial-autocratic, competitive-narcissistic, aggressive-sadistic traits), (4) the medical officer's subjective rating of morale level and his sociometric rating of the preference of the men for one another per quarter, and (5) possible correlates between reported medical symptoms and concurrent social and climatic conditions.

Palmai's results indicate that over the year the men tended to become generally more responsible but less cooperative and conventional. Morale was lowest in the third quarter of the year; during this quarter the incidence of negative social and emotional feelings increased significantly, as did the incidence of somatic symptoms. Being fully aware of the methodological limitations in his work and the tentative nature of his findings, Palmai entered a plea for additional research in these polar 'laboratories' with objective measures by researchers better trained in psychological methods. He particularly emphasized the methodological problems inherent in participant observation, such as observer interference, observer bias, the tendency for the observer to enhance contrasts, and the equivocal nature of any final conclusions. For such research in the Antarctic a very special kind of person, trained for the dual role of participant and observer, is required.

Direct observation by nonparticipants as described by *Radloff* [1973] for saturation diving has not been possible in the Antarctic. But, in the future, polar-orbiting satellites might be applied to this task. A satellite could also ameliorate the severity of isolation by substituting current television programing for the station movies as well as by eliminating communication blackouts and delays. Meanwhile, behavioral science research in Antarctica must continue to rely primarily on indirect observational methods. *Nelson* [1973] has prepared an excellent review of these methods with particular emphasis on the characteristics of isolated and confined groups that merit study, e.g., goal patterns, group structure, group culture, and group effectiveness criteria.

GROUP STRUCTURE AND SOURCES OF GROUP CONFLICT AT SMALL STATIONS

The antarctic station represents a closed social environment, i.e., a total institution. When the men in the party are first removed from the direct influence of their accustomed culture, they not only are spatially isolated but are strangers to one another. They enjoy little privacy and are forced to intermingle with an extremely heterogeneous group. Each man represents, in effect, a separate culture at this time. Moreover, they find themselves burdened with excess leisure time. Common responses to this situation are: (1) depression, loss of hope, and social withdrawal to 'wait out' the winter; (2) subjective exaggeration of the group heterogeneity, polarization into subgroups based on unresolved opinion differences, and deterioration of the party into several hostile cliques; and (3) a period of intense personal social comparison, followed by the evolution of a new homogeneous microculture somewhat different from any of the microcultures originally represented by the individual men and adapted to this special environment. All 'successful' wintering parties at small stations appear to have attained some degree of group homogeneity based on social comparison processes and sociocultural reorganization.

Festinger [1954] formulated a theory of social comparison processes that suggests that the problems and conflicts that arise are only natural consequences of this resocialization process. The greater the differences between the original cultures represented, the greater the social stresses that will develop, and the greater the number of conflicts to be resolved. Our knowledge of how this homogenization occurs is still quite limited because sociologists and anthropologists have been interested mostly in group solidarity and cohesiveness rather than conflict. Knowledge of the day-to-day, week-to-week pattern of conflict resolution within the group is needed.

It is evident that many of the initial attempts of acculturation are often irrational, systematically biased by incorrect judgments, or overly influenced by stereotypes. Some groups may never achieve a stable configuration. The social environment of each station and the microculture that develops appear to change significantly from year to year [*Natani et al.,* 1973]. Both science and individuals suffer when there is no common culture at the station extensive enough and sensitive enough to regulate strong counter motives, promote task accomplishment, harmonize social relationships, and rejuvenate itself whenever the conditions demand.

The U.S. stations have been unique among antarctic stations in that they have been staffed with a split command and two organizational subgroups, one scientific and the other logistic, each with different cultural, occupational, and career orientations. The station scientific programs are conducted by civilians working in the U.S. Antarctic Research Program (USARP). USARP coordinates government and university research services (e.g., map making, specimen sorting, and translation services) under the auspices of the National Science Foundation. Activities of a logistic nature, e.g., transportation, station maintenance, power generation, food preparation, and medical services, are undertaken by the U.S. Navy. *Gunderson and Nelson* [1966*a*] have found sufficient personality differences among different occupations in the logistic (Navy) personnel to indicate that three, not two, distinct subgroups actually exist in antarctic crews: the military administrative-technical personnel (ad-techs), the military skilled tradesmen (Seabees), and the civilian scientists and technicians (USARP's).

At Australian antarctic stations, staffed by civilian scientists and former military personnel, *Law* [1960, p. 275] noted social conflicts between the two subgroups. He stated that, 'On one or two occasions, cliques formed by a group of older ex-servicemen on one side and the young newly-graduated scientists on the other have caused us concern, because of the complete lack of sympathy each has displayed for the ideas of the other.'

Communication between individuals and their various subgroups in a closed antarctic station is often hampered by disparity in cultural levels [*Law,* 1960], differences in occupation [*Gunderson,* 1966*a,* 1968], and pay differences [*McGuire and Tolchin,* 1961]. Moreover, men with strong needs for gregarious activity and achievement tend to suffer more in such an environment [*Gunderson,* 1966*b*] than better-educated isolates who do not become members of cliques, make only a few friends, and occupy their leisure time in elevative activities [*Palmai,* 1963*a*]. *Gunderson* [1966*b*] studied the effects of differences in work load at different times during the winter, a factor that *Law* [1960] and *McGuire and Tolchin* [1961] considered in describing fluctuations in station morale. During periods of low work load, morale tends to drop. It has been noted that the importance of the work role at antarctic stations may result in both striving for perfection and concern over reduced efficiency. It follows then that men who are engaged in satisfying work should be less sensitive about their social relationships than men who are dissatisfied with their work (R. W. Radloff, personal communication, 1971).

The cultural and psychological differences among the various occupational subgroups found at antarctic stations have been explored in considerable detail by *Gunderson* [1966*a, b*]. Studies based on screening tests and field reports suggest that different occupational subgroups are characterized by different motivation and personal values [*Gunderson and Nelson,* 1966*a*], personality traits [*Gunderson and Mahan,* 1966], job satisfaction and performance [*Doll and Gunderson,* 1969*a*], and differential performance predictability [*Doll et al.,* 1969]. *Gunderson* [1966*a*, p. 24] summarized the social consequences of occupational differences with this statement:

> Differences in cultural background characteristics and personality needs tended to be related to incompatibility among work associates during the latter part of the Antarctic winter period under conditions of maximum confinement and isolation from the outside world.

Studies at South Pole station [*Natani,* 1971] support several of the specific characteristics identified by Gunderson et al. in the men of the three occupational subgroups. From self-reports obtained from the men, Gunderson et al. found that the Seabees tended to describe themselves as being excitable with need for support and being conforming and rigid in social relationships, the ad-techs expressed a need for achievement, and the USARP's indicated that they were autonomous and not excitable. Natani's study confirmed that the Seabees were indeed the most excitable of the three subgroups, as was indicated by their wide fluctuations in affective behavior in response to environmental situations such as the isolation of the station for the winter, a midwinter emergency, and preparation for departure at the end of the winter. The USARP subgroup, on the other hand, made essentially no affective response to the first two situations, consistently maintained a higher level of adaptive behavior than the Navy subgroups, and showed only some negative response to preparations for departure. The ad-techs fell between the other two subgroups in their responses to environmental input and occupied themselves reading and studying for promotion examinations (three of the four enlisted ad-techs were promoted to chief petty officer during the winter). In pursuit of their promotions, they attained the highest reading time reported for any of the subgroups.

In general, all three subgroups spent much of their leisure time attending movies, participating in bull sessions, and listening to popular music [*Doll and Gunderson,* 1969*b; Natani,* 1971]. Competitive activities such as chess or checkers, cards, and pool were reported by *Doll and Gunderson* [1969*b*] as being participated in least. They interpreted this finding as being indicative of a need for maintaining harmony. However, Natani found during the 1967 winter at South Pole station that chess, cards, poker, and pool enjoyed popularity throughout the winter and that at times the stakes were abnormally high. For example, a man might shoot a game of pool for 20 cases of beer or roll dice for $100. Possibly, competitive activities, especially those involving gambling, were not faithfully reported by the men in the Doll-Gunderson study owing to the Navy's acknowledged disapproval of such activity.

Natani encountered a problem of this nature when he attempted to get the men to report how much beer and distilled liquor they drank. Some of them feared that they might be classified as alcoholics. The estimated consumption per man including nondrinkers at South Pole for the 1967 winter period was six cases of distilled liquor and 54 cases of beer, suggesting that the needs of modern polar men for alcoholic beverages are much the same as what *Wilkins* [1973] has described them to be for Shackleton's men.

A possible Navy-civilian conflict involving music was identified by *Doll and Gunderson* [1969*b*], who found that the two Navy subgroups tended to prefer country-western recordings, whereas the USARP's leaned toward classical music. The differences in tastes apparently became accentuated as the winter wore on. Differences in musical taste were noted by Natani also, but actually little conflict arose over the issue during the 1967 winter, for each subgroup had its own music-listening area. Moreover, subgroup tastes seemed to drift toward a common denominator until preparations for departure began, when they were reaccentuated. In Natani's experience, Navy-civilian conflicts seem to stem from a more basic problem, that of territorial rights. Problems concerning the mutual use of territory may explain the *Doll and Gunderson* [1969*b*] findings that the Seabees were unique in their avoidance of physical exercise, since all station personnel used the same equipment and exercise area.

In general, the Navy personnel tend to consider the station a Navy facility, feel a sense of ownership for equipment not used daily by the USARP's in their work, and view the USARP's as guests in their facility. At the same time the USARP's feel that the station is a scientific facility and that their work has priority and view the Navy as being on hand to attend to the tasks of station operation and to lend the USARP's assistance as needed. The fact that these men are seldom adequately trained for their antarctic roles and seldom indoctrinated in the psychosocial problems inherent in station life is an obstacle to mutual adaptation. Neither Navy personnel nor USARP's are properly prepared in a psychological sense to normalize differences and develop the necessary synergistic relationships.

One of the major problems, and perhaps the greatest handicap, influencing each new station party is the absence of clearly specified group goals. The scientists are given a brief introduction to the 'big picture' of antarctic research at the annual Skyland orientation conference, but their goals remain primarily individual rather than group or program oriented. The Navy personnel have only a minimal understanding of their 'science support' role, find it difficult to establish a commitment to the USARP mission, and tend to remain devoted to their own individual goals within the routine tasks of station operation and maintenance. This absence of clearly specified group goals for the antarctic station party tends to focus the primary mode of group functioning upon a social-emotional orientation instead of task orientation [*Vallacher et al.,* 1974]. Thus the formation of a group microculture characterized by a wide distribution of friendship choices, attitudes of mutual cooperation, and cohesiveness, which foster positive interpersonal relationships, becomes more important for both efficient task performance and maintenance of individual feelings of well-being.

FACTORS INFLUENCING GROUP COHESIVENESS

Gunderson and Nelson [1966*b*] identified three individual adjustment criteria appearing to influence the end product for both the individual and the wintering-over group and evolving from the initial confrontation of the Navy personnel and the civilians at a small antarctic station, the end product being qualitatively defined as either a 'good' year or a 'bad' year. These criteria are: task motivation, emotional stability, and social compatibility. Gunderson and Nelson outlined the development of these individual performance criteria and discussed the relatively high correlation found between the emotional and the social behavior area. We refer to these individual factors here because they appear to influence the social environment that emerges at a station. The social milieu may have a greater effect upon subsequent individual subjective responses to other individuals than the positions of the men on a compatibility-incompatibility continuum. It has been suggested by others that personality characteristics relevant to task accomplishment directly influence incompatibility in the members of isolated groups [*Haythorn and Altman,* 1967].

Haythorn and Altman conducted an extensive laboratory study on the performance of socially isolated groups as a function of interpersonal compatibility and subjective stress. In their Navy volunteer dyads, personality dimensions such as need achievement, need dominance, need affiliation, and need dogmatism were varied in conjunction with competitive, congruent, and complementary group dimensions. The following six aspects of the results were described in a series of reports: level of intimacy and interpersonal disclosure [*Altman and Haythorn,* 1965], emotional symptomatology related to stress [*Haythorn et al.,* 1966], performance [*Altman and Haythorn,* 1967*b*], personality factors [*Haythorn and Altman,* 1967], social activity and territorial behavior [*Altman and Haythorn,* 1967*a*], and the subjective effects of long-term as opposed to short-term isolation expectations [*Taylor et al.,* 1968]. The findings suggest that territorial behavior and social withdrawal, which increased over time in isolation, are directly related only to the interpersonal traits of dominance and affiliation.

Haythorn [1973] recently reviewed these studies in preparing an integrated appraisal of the results. He concluded that a curvilinear relationship exists between stress and performance and hypothesized that existing closed social systems for urban man generally do not satisfy all higher-level human needs. Meanwhile, *Altman* [1973] used the same studies as source material to develop an ecological approach to the analysis of interpersonal processes in socially isolated groups that emphasizes the use of space and territorial behavior.

Gunderson and Nelson's [1966*a*] data suggest that the Seabees and the USARP's may be incompatible on a need affiliation dimension. Insofar as the Haythorn and Altman results can be extrapolated to antarctic personnel, this factor may partially explain the conflict over territory. Territoriality as related to available personal space also may be a significant factor in the different adjustment levels demonstrated by these two subgroups. The Seabees have the least personal space available to them. They are restricted to their rooms and a work area that is usually shared and not comfortable enough to encourage spending off-duty time there. The cook may lay claim to the galley, but virtually everyone feels that he has a right to invade the galley at any time to prepare a snack or have a card game with friends over coffee. The USARP's, on the other hand, have their sleeping rooms, an office in the science building, and usually one or two other areas such as laboratories, shops, or equipment sites that 'belong' to them. Most of the USARP areas are comfortable enough to encourage extended occupancy and provide varied opportunities for withdrawal from the presence of others using adjacent space.

Another factor that may have a significant impact on Seabee-USARP relationships is suggestibility, a trait not mentioned in connection with the Seabees' personality dimensions. Some support for the contention that the Seabees in general may be hypersuggestible in the Antarctic can be found in the vast literature on sensory deprivation (reviewed by *Wheaton* [1959] and *Suedfeld* [1964, 1969]). Of interest here are findings suggesting that restricted environments facilitate attitude manipulation.

Jones and Goodson [1959] suggested that boredom without sensory deprivation may lead to hypersuggestibility. *Gunderson and Mahan*'s [1966] data indicate that the Seabees are more likely to be bored than the other two subgroups. In another study, *Walters et al.* [1960] obtained evidence indicating that social isolation accompanied by anxiety leads to increased susceptibility to social influence. Finally, if we may infer from *Gunderson and Nelson*'s [1966*a*] data that Seabees are relatively concrete in their thinking, *Suedfeld*'s [1964] findings are of interest. He demonstrated that concrete individuals tended to show enhanced susceptibility to propaganda presented after sensory deprivation, whereas abstract (conceptually complex) individuals were not significantly influenced.

If the Seabees are more open to suggestion under the conditions present at a small inland antarctic station, then it may become possible for a single dogmatic, hostile, and argumentative individual in this subgroup to evoke an anti-USARP attitude on the part of the Seabee group. But an agitator may not be required, since the unusual circumstances under which the Seabees must work may tend to foster feelings of resentment against the USARP group. These conditions are explored in more detail in the remaining discussion in this section.

Pepitone [1964, 1967] described a number of situations fostering hostility and aggression that may apply here. The Seabees' status and security, both based to a large extend on task accomplishment, may be threatened by the intimate presence of personnel with superior formal education who tend to evaluate their performances on dimensions unfamiliar to Seabees. In addition, the uncertainty inherent in coping with the physical hazards of the antarctic environment is difficult to reduce. Seabees who come to feel that they cannot cope with task demands find themselves in a conflict bind. At small stations, usually only one Seabee occupies each job role; hence a man has no one with whom to directly compare his performance. If he fails to perform, his associates may lend assistance, but he has lost face for himself and his subgroup and may feel extremely guilty.

The environment also frustrates the Seabees' sense of task accomplishment. Their understanding of the role of scientific support is often ill conceived, and they have no established standard of performance for this role. Meanwhile, the Antarctic stubbornly resists traditional Seabee efforts at environmental manipulation. *White* [1961] stressed that dealing with the environment implies changing one's relationship to it. The apathy often reported in these men may be due to the absence of 'effectance satisfaction,' which White feels is necessary for sustaining interest in our day-to-day activities. He indicated that men seek to produce effects in the world about them that increase the efficiency of their control over the environment. During the winter the absence of change is ubiquitous, time itself may seem to stand still, and efforts to change local living conditions often unavoidably are met with failure. *Mechanic* [1967] commented that a defense against loss of self-esteem, especially when it arises from unfavorable comparisons with other individuals (as the Seabees may compare themselves to the USARP's), is to come to the conclusion that the others (USARP's) are 'slobs.' The Seabees may then stress and gain gratification from comparisons that make clear their excellence.

The finding by *Gunderson and Mahan* [1966] that incompatibility among work associates is greater during the latter part of the winter than at other times may be due as much to differences in competence as to heterogeneity of cultural backgrounds and personality needs. By midwinter the men who are most likely to contribute to group well-being in the face of danger have been identified. Unfortunately for the USARP's the Seabees may have come to view civilians as liabilities, as being incapable of coping with danger in an emergency. *Pepitone* [1961] suggested that the chief determinant of social behavior in a threatening situation is the training and the ability of the group to withstand or reduce danger. He hypothesized that [*Pepitone,* 1961, p. 9]:

> If the relevant available capacity of a given group to handle a given danger is low or negative compared with other groups, the tendency to affiliate with or develop positive attitudes toward that group will be weak or negative . . . . On the other hand, if the relevant capacity of the group is high compared with other social response contingencies or if the group members are themselves responsible for relatively low danger, the tendency to affiliate with or develop positive attitudes toward that group will be strong.

Mechanic [1962, p. vi] argued that:

> . . . the groups to which a person belongs set the limits, provide the alternatives, and define the meanings to be attributed to threatening as well as non-threatening situations.

Mechanic [1962, p. 220] then made the following point:

> The magnitude of stress is dependent upon the extent of imbalance between the group's environment and the patterns available for dealing with the environment. When imbalances exist between environmental problems and cultural and social resources, we have what we might call 'stress situations inherent within the cultural and social systems.'

From the foregoing information then, we contend that a good year is more likely when the initial USARP-Seabee evaluations of each other remain neutral and when both subgroups as they interact over time demonstrate adequate competence and resourcefulness in dealing with the antarctic environment.

In concluding this section, we think that it is worth noting a possible solution to some of the problems leading to intersubgroup conflict, i.e., training. Both *Wilkins* [1952] and *Wheaton* [1959] stressed that men in isolation who have been poorly indoctrinated and inadequately trained and who are subject to unimaginative administration tend to react with negative behavior.

FIELD DATA COLLECTION TECHNIQES

The methods of observation used by *Natani* [1971] parallel the data collection techniques advocated by *Radloff and Helmreich* [1968] and *Radloff* [1973]. In their study of individual and group reactions to stress in SEALAB 2, Radloff and Helmreich indicated that the closest environmental parallels to SEALAB are space flight and wintering at the south pole. In complex field settings such as these they recommended objective observation and recording of behavior rather than more sophisticated observations requiring extensive interpretation. They also admonished potential investigators collecting data in the field to refrain from asking their subjects questions. The SEALAB 2 report describes well the problems, planning, and failures attending psychosocial research in real life situations. Controlling and manipulating variables in the field are extremely difficult tasks, and chance events often lead to failures or to unexpected opportunities. In the SEALAB 2 study, equipment failure prevented the collection of intragroup communication data. Radloff and Helmreich considered this one of their most crucial sources of potential data for the determination of adjustment, group interactions, leader dependence, group cohesiveness, and intersubgroup differences. (*Chapple* [1940] suggested the systematic measurement of interaction as an element of social behavior.)

On the other hand, in Natani's South Pole study (in the field in 1967), an unscheduled midwinter station evacuation (May 16, 1967) greatly added to the value of group interaction data. It provided an opportunity to observe individual responses to enhanced stress against a background of chronic stress. Although the station evacuation was relatively short in duration (24 hours), its stress-provoking qualities were prolonged and exacerbated by the effects on the men of the news that U.N. forces had withdrawn from the Middle East on May 19 and later, in June, that the Arabs and the Israelis were waging war. At that time the men actively contemplated the possibility of being abandoned because of another major world conflict. The results of Natani's study, then, must be interpreted on two levels. One concerns the behavior expected under normal conditions at the station, and the other concerns the influence of unexpected stress on the familiar conditions. Specific aspects of this event will be discussed in the section on the transient effects of the station emergency.

PARTICIPANT OBSERVATION AT 90° SOUTH

The primary intent of Natani's study was to obtain measures of behavior indicating levels of adaptation to the restricted environment of an antarctic station. One independent variable considered in analyzing the data was the amount of time that the men spent at the station. Once the last summer flight had departed in February, every man was committed to spend at least 8 more months at South Pole station. The dependent variable chosen for measurement was time use. Two methods of obtaining measures of time use were employed: self-reports of their activities, obtained from the men on three separate occasions at spaced intervals (W1, winter 1; W2, winter 2; and W3, winter 3) during the winter on activity cards (Figure 1), and continuous, unobtrusive observation of their leisure time activities (Table 1). Data from both these measures were analyzed separately for occupational subgroup differences in time use and changes in time use as the winter progressed. Leisure time was defined as off-duty time. With respect to leisure time the definition was the same as that stated by *Jennings* [1950, pp. 229–230]:

> Leisure time, to the population under study, comprised time *for being with* whatever persons the individual wanted to seek out; or he could, if he wished, spend all or much of such time alone . . . . Thus *leisure,* as here designated as a criterion for sociometric choice, means in choice expression the individual's decision for or against simply being with particular other members of the population in the time at his disposal, which could be filled as the individual saw fit, according to his inclinations . . . .

The difficulty of obtaining sociometric choices at antarctic stations was noted earlier. As *Radloff* [1973] has indicated, negative sociometric choices are especially difficult to solicit; therefore only positive choices were recorded for social interactions. This technique was considered to give a valid measure of positive choices, because Natani's experience substantiated the assertion of *Sells and Gunderson* [1972] that men in a closed environment tend to avoid conflict by withdrawing from face-to-face contacts.

The definition used for a social interaction was obtained from *Bales'* [1950, p. 33] description of a small group:

> [An instance] . . . in which each member [present] receives some impression or perception of each other member distinct enough so that he can, either at the time or in later questioning, give some reaction to each of the others as an individual person, even though it be only to recall that the other was present.

The self-report sleep cards, described by *Lewis and Masterton* [1957], were used as a model in preparing a card in IBM format (Figure 1) for the collection of self-reports in half-hour intervals of the time that men spent in both sleep and common leisure activities. An IBM card format was selected for several reasons: (1) the size of the card made it easy for a man to carry it all day in a shirt pocket; (2) a format could be developed to facilitate data reduction, possibly keypunching directly from the card; and (3) it was assumed that the indoctrination of Americans not to 'fold, spindle, or mutilate' such cards (implying a certain sacredness of IBM cards) has become very effective in our society. The self-report cards were initially field-tested at South Pole during a summer study of sleep and activity patterns [*Shurley et al.,* 1970]. During the summer study, methods of distributing and collecting the cards were developed. Clarification of equivocal reports did not prove difficult, and the results of the summer study indicate that a high degree of uniformity of response in terms of categories was maintained and that the data obtained are discriminating and reliable.

Three times during the winter, spaced at approxi-

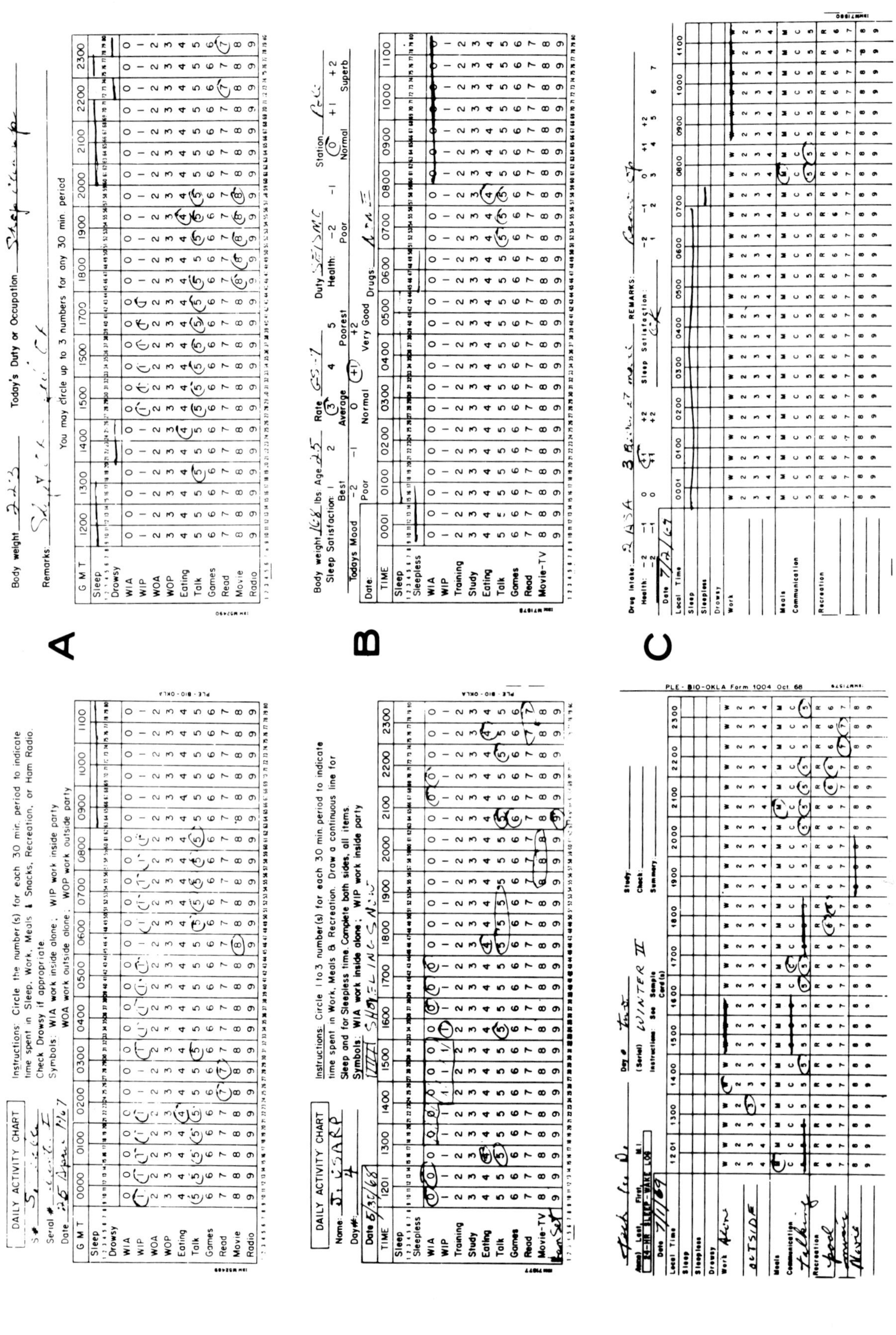

Fig. 1. Front and back samples of a keypunch card developed to obtain daily self-reports of sleep and waking activity in half-hour increments as well as subjective global ratings of sleep satisfaction, health, and mood. Form A is an initial version used by Natani in 1967; form B is a slightly modified version used by A. Joern in 1968; form C is the format currently used by the Behavioral Sciences Laboratories of the Oklahoma City Veterans Hospital.

TABLE 1. Adaptation Rating Scale

Positive (+1) Behavior	Negative (−1) Behavior
1. The individual was alone by choice when opportunities for social interaction were available.	1. The individual sought social interaction and was not observed performing solitary activities.
2. The individual participated in a variety of game activities.	2. The individual was observed participating in only one type of game activity or in no game activities.
3. The individual spent a portion of his leisure time reading.	3. The individual was not observed reading.
4. The individual performed extra work during his leisure time.	4. The individual was not observed working during his leisure time.
5. The individual's observed frequency of social interaction was below the mean frequency for his occupational subgroup.	5. The individual's observed frequency of social interaction was above his subgroup's mean.
6. The individual was not observed consuming alcoholic beverages.	6. The individual was observed consuming alcoholic beverages.
7. The individual was not observed taking snacks at odd hours.	7. The individual was observed taking snacks at odd hours.
8. The individual was not observed discussing symptoms or making subjective complaints.	8. The individual was observed discussing symptoms and/or making subjective complaints.
9. The individual's observed frequency of intersubgroup social interaction was above the mean for his occupational subgroup.	9. The individual's observed frequency of intersubgroup social interaction was below his occupational subgroup's mean.
10. The individual participated in an exercise program.	10. The individual was not observed exercising.
11. The individual played an active part in group discussions (bull sessions).	11. The individual was not observed participating in group discussions.

mately 2-month intervals, the men submitted daily self-reports of their activities for 1 week. The data points used were W1 (April 24 through April 30), W2 (June 26 through July 2), and W3 (August 14 through August 20) (Table 3 and Figure 4). Blank cards were available at all times during the data collection weeks in a dispenser mounted in the station galley. Each morning the investigator collected cards from the previous day and immediately checked them for omissions or ambiguous reports, which were then clarified with the men while the reported activities were still fresh in their memory.

In general, all the men cooperated well. However, data from only 19 of the 21 subjects are reported because of some reluctance on the part of two men about filling in and returning their cards on a daily basis. Consequently, the reports that they submitted were not considered sufficiently valid for use in the study.

The unobtrusive observations of leisure activities between 10 P.M. and 8 A.M. made by Natani were used to determine scores on a rating scale of adaptation. This bipolar scale consisted of 11 items indicating positive or negative adjustment (Table 1). The activities covered by the scale included choice of leisure activities, social drinking behavior, eating behavior, vocalized subjective complaints, participation in physical exercise, and frequency of social interaction. The total score on the scale for each subject could theoretically range from +11 (excellent adaptation) to −11 (very poor adaptation). It was hypothesized that the score that an individual obtained on the rating scale could be interpreted also in terms of affect (i.e., a negative score indicating unpleasant feelings and a desire to leave the environment).

Scale scores were computed monthly, and the observations taken during the month were reduced to binary quantities. The actual frequency with which an activity was observed during the month was disregarded, and the individual's score for the month for each item was determined by the presence or absence of observations in a particular activity category. Binary reduction was considered necessary to compensate for methodological problems, such as the tendency of an observer to enhance contrasts and harbor biases. The resulting scale was an ordinal scale, and therefore the net scores derived from the various positive and negative scale items may not have given a true indication of the relationships involved because of the nonlinear nature of ordinal scales. Nevertheless, after these scores were normalized by adding a positive constant to each score, a parametric analysis of variance (Anova) was performed on them. An Anova was performed because the scale range (23 points) was great enough to consider the scale approximately interval, and a normal distribution of scores was obtained (Figure 2). The slight negative skew of the distribution in Figure 2 may be an index of the negative feeling often associated with monotonous, unvarying, highly predictable conditions [*Fiske and Maddi,* 1961].

The greatest variance among the occupational subgroups appeared in March and May, and the lowest points on the scale occurred in March and October (Figure 3). Throughout the period of observation, little variation occurred in the USARP group. March and October may have been atypical data periods, however, in which different social processes were operating from those observed between April and September. There are several reasons that these months could have been atypical: (1) sunset occurred in March, and sunrise occurred

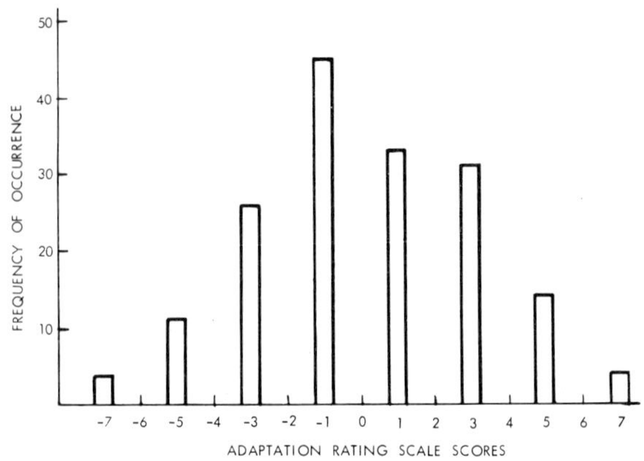

Fig. 2. The frequency distribution of raw adaptation rating scale scores derived from specified behavioral observations of 21 men each day from 10 P.M. to 8 A.M. during 8 months of isolated duty at South Pole station.

in October; (2) the work loads in March and October were greater and involved more outside activity, especially for the Seabees; (3) the men were still evaluating one another in March, and in October they were anticipating departure and in many instances never expected to see their wintering-over companions again; and (4) the observer was not as experienced in March, nor could observations be made as systematically in October, when men were again standing night watches. Therefore the March and October data points were deleted from this analysis. Subgroup differences were found to be nonsignificant in the analysis of the winter scores.

The total variation across months was significant ($p < 0.05$) but could not be accounted for by linear, quadratic, or cubic trends. The obviously discrepant mean obtained in May (Figure 4) differed significantly (t = 2.54 and $p < 0.02$ by the repeated measure t test [*Walker and Lev,* 1953]) from the June mean. Evidently, the majority of the variance found in the adaptation level scores may be attributed to the May mean. The implication is that, if adaptation occurred during the dark months, the rating scale was insensitive to it, even though the scale seems to be sensitive to stress. The stress sensitivity of the scale was supported by a reexamination of the raw data and environmental events in May. On May 16 the entire station was evacuated for 24 hours in -67°C (-88.6°F) weather when an 18.93-liter (5 gal.) drum of carbon tetrachloride was accidentally broken open and the highly toxic fumes permeated the station atmosphere. From May 1 through May 15, positive rating scale criteria were twice as numerous as negative criteria. From May 17 through May 31, negative rating scale criteria exceeded positive criteria. The scale items most augmented by and thus apparently most sensitive to stress were the frequency with which men stayed up past midnight and the frequency of social drinking. During this period when they went to bed late their extra time awake was spent drinking, watching movies, or playing cards. Bull sessions per se, without alcohol consumption, almost ceased to occur during the late hours.

The only indication of different responses to the emergency by each group was found in the social interaction data (Table 2). In May the ad-tech intrasubgroup interactions diminished sharply, and Seabee intrasubgroup interactions almost doubled, whereas the USARP intrasubgroup interactions remained essentially unchanged. The Seabee subgroup showed the greatest variability in intrasubgroup interactions.

RELATIVE FREQUENCY OF REPORTED LEISURE ACTIVITIES

Both *Natani* [1971] and E. K. E. Gunderson (personal communication, 1970) obtained self-report activity data from the same group of men wintering at South Pole in 1967. The relative frequency (RF) of those activities reported with the highest frequency during the winter was calculated for both sets of data by Natani and compared with the RF of the leisure time activity reported by 80 temperate zone aerospace employees [*Eddowes,* 1961]. These data (Table 3) indicate that talking is consistently the most popular leisure time activity for the men at South Pole. The overall average reported at the three winter data points (W1, W2, and W3) is approximately 4.5 hr/day. In comparison with data compiled by *Eberhard* [1967], this value is 9 times greater than the social talking time usually reported by temperate zone men. Eberhard found that time expended in conversation per se by the U.S. population as well as the populations of nine other western nations amounts to

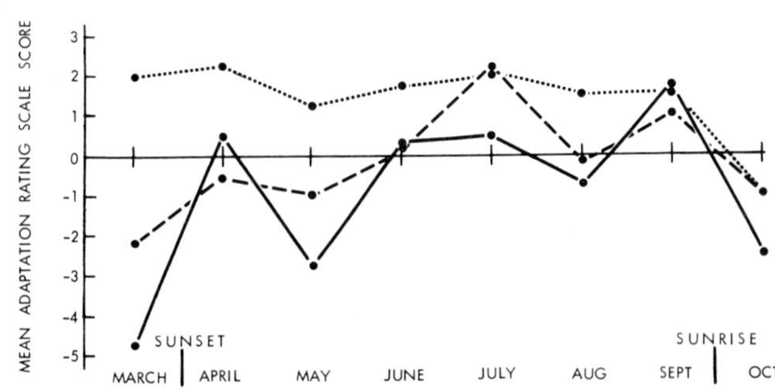

Fig. 3. Comparison of raw mean adaptation rating scale scores derived from specified behavioral observations of the members of three different occupational subgroups during 8 months of isolated duty at South Pole station. Solid lines indicate Seabees (N = 8), broken lines indicate ad-techs (N = 5), and dotted lines indicate USARP's (N = 8).

only 0.2–0.5 hr/day. This large difference in time use may be due to the use of incompatible methodological techniques in the collection of time use behavior.

A temperate zone comparison study by W. L. Fesler and J. T. Shurley (unpublished data, 1967) of urban industrial workers, using the same activity card instrument as was used at South Pole, indicates that men at South Pole spend twice as much time engaged in social talking. The South Pole data also suggest that the Seabees are the most verbose group. The temperate zone population studied by E. E. Eddowes (aerospace employees who probably resemble the USARP's) chosen for comparison here may have exhibited a high degree of taciturnity, since a summation of the RF of such items as social and family activities, where talking is to be expected, falls far below a summation of the RF of these activities in Antarctica. An alternative conclusion is that some facet of the station environment stimulates vocal behavior.

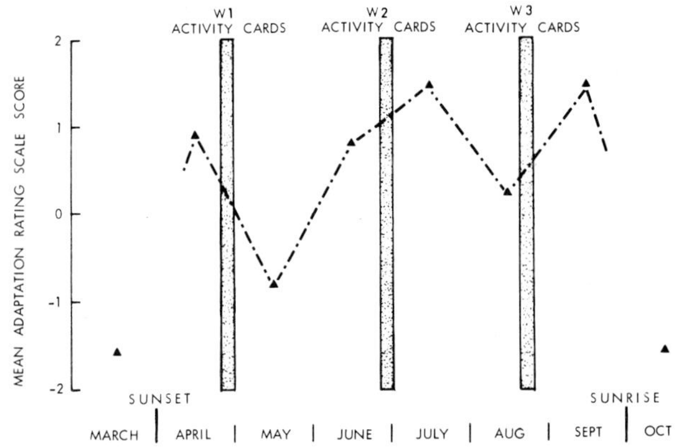

Fig. 4. Variation in the group (N = 21) raw mean adaptation rating scale score derived from specified behavioral observations of the men each day from 10 P.M. to 8 A.M. during 8 months of isolation. The three vertical bars indicate the weeks during which time use information was collected from the men on self-report daily activity cards.

SOCIAL COMPARISON PROCESSES IN THE ANTARCTIC

In his review of off-duty time use by confined groups, *Eberhard* [1967, p. 3] indicated that men in confinement spend twice as much time eating as men in the general population. He also noted that men at South Pole spend almost twice as much time eating during the winter as they do during the summer. *Eberhard*'s [1967, p. 41] statement concerning Antarctica was based largely on preliminary data collected by Natani, and it now appears that the extra eating time noted is evidence that the men took advantage of mealtimes to linger over their coffee in conversation. In Natani's study the amount of time spent talking far exceeded expectations for time use. Talking activity was analyzed under three categories: net social talking, talking during meals, and talking during game playing. The Seabees talked more ($p < 0.05$) while they played games than the USARP subgroup. The other talking categories were nonsignificant across the subgroups.

Altman and Haythorn [1965] in their dyad isolation study found that social isolation affects patterns of interpersonal exchange in a way that leads to more intimate self-disclosure. They hypothesized that, to manage close mutual involvement in isolation, the subjects learned more about each other. They also suggested that the other person may have become the only source of varied stimulation. These factors probably contributed to the marked increase in interpersonal exchange found at South Pole, but the main motivation appears to be social comparison [*Festinger,* 1954; *Schachter,* 1959; *Mechanic,* 1962].

Both *Schachter* [1959] and *Weller* [1963] found that anxiety arouses a desire to be with others experiencing similar problems because they serve a direct anxiety-reducing function. Both Schachter and *Mechanic* [1962] identified essentially similar factors in ambiguous stress-provoking situations that lead to social interaction and social evaluation. When objective criteria are absent, individuals in groups under stress have been found to use social comparisons to evaluate their own feelings of

TABLE 2. Mean Number of Intrasubgroup and Intersubgroup Interactions Observed between 10 P.M. and 8 A.M.

Interaction	N	March	April	May	June	July	August	September	October
				Intrasubgroup					
Ad-tech/ad-tech	5	5.0	12.0	4.4	10.4	8.0	13.2	9.2	8.8
Seabee/Seabee	8	5.0	18.0	32.1	18.5	16.5	27.7	33.5	10.5
USARP/USARP	8	5.6	19.0	18.5	21.0	19.0	28.1	25.2	21.5
				Intersubgroup					
Ad-tech/Seabee	13	2.46	7.5	6.3	8.4	5.9	9.0	9.0	6.0
Ad-tech/USARP	13	2.76	8.76	4.07	10.5	6.7	9.0	9.3	7.0
Seabee/USARP	16	1.37	6.1	5.75	9.4	7.37	11.2	9.6	6.75

TABLE 3. Comparison of the Relative Frequency of Specified Leisure Time Activities Obtained from Five Different Self-Reports from South Pole and One Report from Temperate Zone Men

Activity	Self-Report Activity Card* W1	W2	W3	Self-Report Activity Data† Early Winter	Late Winter	Data on Temperate Zone Men§
Talking	0.964	0.981	0.976	0.933	0.809	0.125 (social activity)
Movie watching	0.863	0.860	0.828	0.819	0.933	0.325 (television, movies, plays)
Games	0.759	0.754	0.688	0.952	0.905	0.676 (composite recreation)
Reading	0.609	0.762	0.666	0.590	0.695	0.725
Radio operation‖	0.172	...	0.151	0.271	0.276	0.100 (family activities)

*The RF (self-report activity cards) equals the total number of times that the activity was reported for 1 week divided by the total number of times that the activity would be reported if each man reported participating in the activity each day of the week.

†The RF (E. K. E. Gunderson, personal communication, 1970) equals the number of men reporting participation in an activity every day and those reporting a few times per week divided by the total number of men reporting (early winter, $N = 17$; late winter, $N = 19$).

§The RF is based on comparable activities for talk and ham patch taken from a study by *Eddowes* [1961]. Watching movies is a composite of watching television (0.300) and movies and plays (0.025). Games include musical activities (0.275), manual activities (0.213), photography (0.100), chess (0.063), and all others (0.025).

‖During W2 the ham radio was inoperative. A ham patch is a 2-way radio-telephone conversation between an antarctic volunteer and usually a close friend, sweetheart, relative, or supervisor in the United States.

fatigue, annoyance, satiation, performance, and appropriate reactions to novel stimuli.

Conversation is one of the most effective and pleasant means of making social comparisons. At South Pole station the men consistently spent most of their talking time discussing those areas of greatest concern to them, i.e., their job (past, present, future), sex, and politics (both within their own immediate administrative hierarchy and of world events relating to international tensions). Therefore the greater incidence of talking reported by the Seabees may indicate that they were more anxious and thus sought company and social comparison more frequently. Talking behavior did decline to some extent during the winter but not significantly. One might infer that this continuing high talk level is another indication of poor adaptation. If the men had improved significantly in their performance or had attained greater inner security, their talking behavior should also have declined significantly.

TRANSIENT EFFECTS OF THE STATION EMERGENCY OF MAY 16, 1967, ON GROUP COMPATIBILITY AND ACTIVITIES INDICATING ADAPTATION

The behavior recorded after the emergency evacuation of the station on May 16, 1967, suggests support for the relationships between social interaction, social comparison, and talk outlined above. During May, when the incident occurred, intersubgroup interactions showed little variation. Intrasubgroup interactions, on the other hand, were affected. The USARP interactions were unchanged, the ad-techs spent less time together, and the Seabees almost doubled their intrasubgroup interactions. The event may have been particularly stressful to the Seabees because emergency quarters and camp evacuation were their direct responsibility. It was soon evident that the emergency quarters were not ready for immediate occupancy and required over 4 hours of preparation before they could be occupied. The evacuation was executed in an orderly fashion and served as an excellent drill, but the Seabees were left with the thought that, had this been a hasty evacuation in the face of fire, the survival value of the emergency quarters would have been greatly reduced. The incident thus reflected directly on their area of task accomplishment. The ensuing increased intrasubgroup interaction is consistent with the social comparison findings of *Schachter* [1959] and those findings relating to cohesiveness and rejection under stress described by *Schachter* [1959] and *Weller* [1963]. During their period of increased intrasubgroup interaction after the evacuation the Seabees were apparently normalizing their individual responses as a Seabee to the incident and evaluating what it should mean to them as a group. *Radloff and Helmreich* [1968] found support for the view that social interaction is related to successful adaptation in an extreme environment. They indicated that, during interaction, social comparisons and emotional ventilations occur that are probably of great value in aiding men in their adjustment to stress.

By June the intrasubgroup interactions for all subgroups had returned to the levels recorded in April. Perhaps of greater interest than subgroup responses is the manner in which the event influenced subsequent behavior for the station complement as a whole. In general, the men expressed a keener awareness of the environment and their dependence on one another for survival. Also, as was noted earlier, they were more sensitive to the influence that world tensions might have on their being relieved at the end of the winter. Although their feelings of insecurity may have increased, most of the men expressed the view that the incident created an enjoyable change. They were more confident in their ability to work together, a feeling of excitement was present, and conversations were often devoted to the problems that might be encountered in 'walking out' when the sun returned.

Station activities were more diverse during June and July. On the positive side, more men engaged in the exercise programs, topics of conversation showed greater variability, reported talking time tended to decrease slightly, and sleep patterns recorded at that time [*Natani et al.*, 1970] significantly shifted toward their preantarctic base line values. However, some negative aspects of behavior that are indicative of stress and thus may reflect increased stress tended to increase, e.g., the incidence of somatic symptoms other than insomnia and the frequency of social drinking. In general, the overall response of the men at that time tends to support the conclusions of *Altman and Haythorn* [1967*b*] concerning performance. They found that performance on tasks demanding cooperation tends to be enhanced in isolation and concluded that performance outcomes depend on the nature and characteristics of intervening events as well as on initial conditions. The evacuation incident suggests that change, even a stressful change, contributes to the positive aspects of adaptation if it raises the men's self-confidence and fosters group cooperation.

The scale suggests that another low level of adaptation was reached in August, although it is statistically nonsignificant. By August the men were rereading their past summer's mail, the liquor supply was running low, all the movies had been seen, the darkness was oppressive, and the temperature plummeted to −76°C (−105°F). Under these conditions some of the men began spending an hour or two each day sitting under sun lamps even though they did not tan [*Natani and Shurley*, 1973], and others started going outside everyday to catch a first glimpse of the sun. With the return of the sun in late September the adaptation level rose again, and there was a general decline in the incidence of somatic symptoms, but the group now drifted further apart as the men began preparations for departure.

OCCUPATIONAL SUBGROUP DIFFERENCES IN SELF-REPORTED TIME USE FOR SIX SPECIFIED ACTIVITIES

The self-report activity data collected by *Natani* [1971] provide support for the hypothesis that civilian and Navy personnel may be identified or characterized by the way in which they use their time in the Antarctic. These activity results appear consistent with the findings of *Gunderson and Nelson* [1966*a*], which indicate that there are basic differences in cultural and psychological attributes among the occupational subgroups wintering at antarctic stations. An Anova indicated that significant differences in time use existed between the three subgroups. Overall subgroup differences for each activity category were then evaluated with post hoc tests in a multiple comparison procedure using Student's t statistic for small unequal n. The t test results are summarized in Table 4. Table 4 shows that the significant differences found for working, reading, and movie watching indicate that the Seabee subgroup contributed the greatest source of variance. No significant differences were found between the subgroups for time reported spent eating, playing games, or talking socially. Work is the only activity for which a significant difference ($p < 0.05$) was found between the ad-techs and the USARP's. The USARP's consistently reported the most work and differed from the Seabees in this activity at the $p < 0.01$ level. The Seabees reported the least amount of time spent in reading but differed less in this respect from the USARP's ($p < 0.05$) than they did from the ad-tech ($p <$

TABLE 4. Summary t Table of Overall Occupational Subgroup Differences for Six Specified Activity Categories

Activity	Group Comparison	d.f.	t	p
Work	Ad-tech to Seabee	9	0.261	NS
	Ad-tech to USARP	10	2.56	<0.05
	USARP to Seabee	13	3.79	<0.01
Eating	Ad-tech to Seabee	9	1.03	NS
	Ad-tech to USARP	10	0.63	NS
	USARP to Seabee	13	0.41	NS
Games	Ad-tech to Seabee	9	0.45	NS
	Ad-tech to USARP	10	0.74	NS
	USARP to Seabee	13	1.62	NS
Reading	Ad-tech to Seabee	9	3.46	<0.01
	Ad-tech to USARP	10	0.52	NS
	USARP to Seabee	13	3.12	<0.05
Movie watching	Ad-tech to Seabee	9	2.25	NS
	Ad-tech to USARP	10	0.21	NS
	USARP to Seabee	13	3.50	<0.01
Net social talk	Ad-tech to Seabee	9	0.31	NS
	Ad-tech to USARP	10	0.44	NS
	USARP to Seabee	13	0.94	NS

Symbol NS means not significant.

0.01) subgroup. The Seabee subgroup also spent considerably more time movie watching ($p < 0.01$) than the USARP's. The general pattern suggests that the Seabee and USARP subgroups often showed divergence in their choice of time use, the ad-tech group falling between the two.

Figure 5 illustrates the variance of the Seabee subgroup across time with respect to the other two subgroups. In April the Seabees worked and watched movies about the same amount of time as the other men but read much less than the USARP's ($p < 0.01$). Although the reading differences between these two subgroups became nonsignificant in June and August, they then differed greatly with respect to time spent working ($p < 0.01$). All three subgroups spent essentially equal time watching movies in April, but the Seabees spent much more time at the movies in June ($p < 0.01$). Even though all the movies had been shown by July, the time reported spent at the movies rose for all the men in August. The August Seabee movie-viewing time was still significantly higher ($p < 0.05$) than the August USARP movie time.

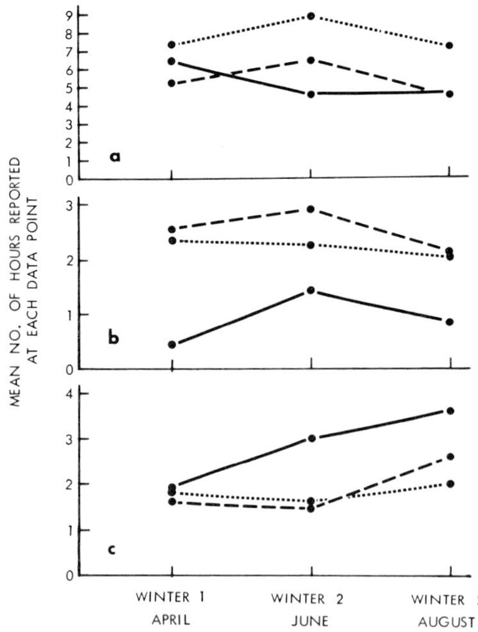

Fig. 5. Comparison of the mean amount of time reported spent per day in the specified activities of (*a*) working, (*b*) reading, and (*c*) movie watching at each data point by the members of different occupational subgroups. Solid lines indicate Seabees ($N = 7$), broken lines indicate ad-techs ($N = 4$), and dotted lines indicate USARP's ($N = 8$).

CONCLUSIONS CONCERNING ADAPTATION TO ANTARCTIC STATION LIFE

Except for the events of May and their brief positive aftereffects at South Pole station the data obtained by Natani support findings that indicate that little psychological adaptation occurs during the winter and that the men maintain, at best, their initial level of adaptation. *Rohrer* [1961] described the period between sunset and sunrise as one of depression with little change, and he noted evidence of regressive behavior instead of adaptation. *Nelson* [1964] investigated group structure with respect to work, formal communication, and off-duty friendships for the men in a number of antarctic stations and concluded that there were no changes during the winter. However, the mere lack of change in adaptation level found in the Antarctic during the winter may not be an abnormal response to a stable, albeit stressful, environment. In a Stateside study, *Bradburn and Caplovitz* [1965] found a strong relationship between feelings reported at the beginning of their study and those reported at the end of their study, the high consistency of the responses indicating that little change occurred over an 8-month period. Those individuals who reported changing from extreme feeling states were more likely to report a small change rather than a large one.

The factors that Bradburn and Caplovitz found to be most significantly related to higher positive feelings in the continental United States are a high level of social interaction and participation in the environment. These findings suggest that the high level of talking time found in the Antarctic may be a measure of positive feelings as well as an indication of stress. But there is always the possibility that intense social interaction in a closed, isolated environment may have a paradoxical effect. In another report, *Bradburn* [1969] cautioned that his measures do not discriminate between different levels of social interaction, nor do they establish a simple causal connection between social participation and positive affect. He indicated that there might be some point beyond which a higher level of social interaction would not be related to positive affect. In addition, note that *Bradburn* [1969] has revised his model and introduced novelty as a component of social participation. Thus the concept of social participation is now viewed as being composed of two analytically distinct components that may interact in producing positive affect. Novelty in this context depends primarily on a wide range of acquaintances, new experiences in different situations, and extended movement in both physical and psychological space. These are the very features that are most prominent among the deprivations suffered by the antarctic volunteer.

Palmai [1963*a*] reported that social withdrawal is

associated with positive adjustment in Antarctica. Social withdrawal may be related to the observation made by *Altman and Haythorn* [1965] that, when isolation subjects make exploratory probes into intimate personal information over time, the high revealers come to dislike one another. However, *Priestley's* [1969] anecdotal account of his 1912 experience as a member of Scott's northern party 'caught out' on Inexpressible Island contradicts their observation for extended antarctic isolation and close confinement.

As we indicated earlier, anxiety reduction via social comparison may be the primary variable influencing socialization at an antarctic station.

A recent analysis of three South Pole diaries suggests support for Bradburn's view that the novelty involved in diverse social contracts is strongly related to positive affect. *Natani et al.* [1973] hypothesized that, when anxiety levels are low, occupational intrasubgroup interactions should be lower, whereas the variability in social interactions in general should increase. This condition would be especially true in an isolated group in which the members have come to rely on individual differences as sources of stimulation, although they might also be avoiding personal overexposure. Data on personal conflicts and subgroup interactions were obtained from diaries, covering a 3-year period at South Pole station, that were kept by the biomedical technicians conducting the University of Oklahoma sleep studies (1967 and 1968) and the medical officer (1969). The results indicate that there were fewer conflicts during the year identified as the best year of the 3 years and significantly greater variability (one-tailed F test) in social interactions during the best year. These findings suggest that high morale during the winter at South Pole is associated with greater variability in social interactions.

Education and cognitive style also appear to be significant factors influencing psychological adaptation to the Antarctic environment. The USARP's have the most formal education, the Seabees have the least, and the ad-techs have a middle amount. This pattern of occupational subgroup relationships was reflected in both rated adaptation levels and self-reports of time use. These results support the *Bradburn and Caplovitz* [1965] finding that 'happiness' is positively correlated with education and income.

Thus there seems to be some support in the antarctic data for the measurement of psychological well-being as advocated by Bradburn and Caplovitz. However, the most significant aspect of their work may lie in the separation of positive and negative affect into two independent dimensions. If they are truly independent, the subjective reports and objective findings of positive personality changes during antarctic wintering in the face of a great variety of negative symptoms and effects are no longer a paradox and should encourage the development of measures to assess positive affect in this environment and other high-risk environments as well as in the life of urban man.

The evidence reviewed here suggests that the USARP's and the Seabees differ significantly in their psychological responses to the antarctic environment, although both may come to express positive regard for the experience. The USARP's began their antarctic sojourn at a positive level of rated adaptation and maintained this level until sunrise and the initiation of departure preparations. The Seabees, on the other hand, began the winter at a neutral level and showed considerable variability in their rated adaptation level from month to month.

Oken [1962] found that normal subjects who demonstrated high affective lability tended to experience greater physiological arousal to stress and also exhibited a greater defensive response. The question may then be asked, Do the Seabees differ from the USARP's in their physiological responses? For the moment any answer to the question must be deferred owing to insufficient knowledge of the psychological and physiological consequences of chronic stress. However, we are reminded that social psychology can be defined as both a social and a biological discipline [*Shapiro and Schwartz,* 1970]. The application of psychophysiological techniques to the study of human social behavior provides the investigator with nonverbal, objective, and relatively bias-free indexes of human responses. The study of changes in sleep patterns as a response to conditions at South Pole by *Shurley* [1971] and co-workers illustrates the use of the psychobiological approach to the study of behavior in Antarctica. The results of these studies indicate that anxiety and arousal remain at a high level all winter.

Another feature of life on the Polar Plateau identified by the psychophysiological study of sleep at South Pole station is chronic hypobaric hypoxia, which may significantly influence human behavior in this environment. The pressure altitude at South Pole is approximately 3353 meters (11,000 feet), whereas at Plateau station, which was in operation from 1966 through 1968, it is approximately 4050 meters (13,500 feet). This additional physiological stress appears to have some interesting psychological correlates.

In comparing positive personality changes in the men at New Zealand's Scott base (at sea level) with those in the men at South Pole, *Taylor and Shurley* [1971] found greater changes, as measured by the R. B. Cattell 16 personality factor test, at South Pole. Natani and A.

B. Blackburn administered the Minnesota multiphasic personality inventory (MMPI) to men at South Pole and Plateau in 1967 on two occasions, early winter and late winter. K. Natani found no significant changes in the profiles of individual men or between occupational subgroups. But *Blackburn*'s [1973] data indicate a significant reduction in verbally expressed anxiety and revealed some differences in occupational subgroup responses. These findings suggest that a relatively high level of psychological and physiological stress is required to induce personality changes, either positive or negative, in healthy men and that the MMPI is not sensitive to the effects of intermediate levels of stress in men who volunteer for antarctic duty and characteristically have initial MMPI profiles suggestive of health and strength.

The most potent sources of stress identified at South Pole are extreme cold, hypobaric hypoxia, markedly different light-dark cycles, lack of novelty, monotonous activity schedules, and 8 months of unbroken group social isolation with close confinement. Common responses to these stressors have been depression, social withdrawal, exaggeration of group heterogeneity, and intense social comparison. The harshness of the environment is reflected in the fact that the most popular choices of leisure activities have been those that are sedentary in nature (e.g., bull sessions, movie watching, and reading). Talking has consistently been the most popular activity and is indicative of increased social interaction. It is concluded that, in this setting, social interaction provides opportunities for social comparison and social evaluation that serve direct anxiety-reducing functions and inevitably lead to the formation of a new microculture adapted to this special environment. Daily observations of the frequency with which men stayed up late at night, the frequency of their social drinking, and the frequency of their intrasubgroup interactions indicate that these behaviors are valid measures of subjective response to stress. These measures suggest that the beginning and the termination of antarctic isolation are the most stressful occasions.

Task motivation, emotional stability, and social compatibility are the primary factors determining the quality of life at antarctic stations during wintering duty. Both the emotional area and the social area need sharper delineation in measurement and definition [*Gunderson and Nelson,* 1966*b*]. The application of psychophysiological techniques seems to offer the best approach to further the study of emotional responses to the antarctic environment. At the same time, the problems of social adjustment continue to challenge the investigator interested in the unique acculturation occurring each year at antarctic stations. This process appears to favor men who are reserved, controlled, diffident but agreeable, and above average in intelligence and socioeconomic status. Instruments sensitive to the positive aspects of social interaction in confinement are obviously needed, and, prior to antarctic duty, some type of training in sensitivity to interpersonal responses and group behavior such as that described by *Rioch* [1970] may be useful in ameliorating to some extent the stresses produced by interpersonal processes and inadequate leadership.

TERMINATION OF ISOLATION

During the long antarctic winter a collection of heterogeneous strangers has formed a group with a distinct microculture that has developed its own measures of control, adaptation, and change with little assistance from outside authority. We hypothesize that, to accomplish this feat, the independent individuals who volunteer for antarctic duty also have developed a strong need for behaviors that give them positive feedback about their own self-control, control over their fate, and control over the environment. The isolates may thus individually and as a group establish internal routines and external expectations that seem idiosyncratic and immature or unrealistic to an outsider but that are extremely important to the station party for structuring time, maintaining self-identities, and providing social security.

In this situation, internal attempts at control by unskilled leaders will be resented and may become highly disruptive. The exercise of authority in the control of activities, such as specifying the time when individuals may eat, drink, work, play, sleep, and communicate with the outside world, may present special problems if it is inconsistent with norms already established by the group. External authority also will be resented because it may have been exercised only rarely in the past, the need for it may not be obvious to the group, and the action requested may not satisfy the group's conception of what is proper under the circumstances.

In addition, both past instances during the year of seemingly undue delays in response to requests for outside action and responses interpreted to be ill considered or inappropriate, such as serious responses to jokes or pranks, will enhance a natural resentment to what appears to be an unfeeling and unperceptive outside authority. The more severe the isolation, the more likely it will be that attempts at control will be resented, and the greater will be the critical concern by the isolates over the empathy of their outside superiors and their competence at providing support.

Those 'back home' who have occasion to deal with the affairs of isolated men should bear in mind that anxiety levels prior to the termination of isolation are high because the group has begun to disintegrate as the individuals begin both physical and psychological preparation to take their separate paths and reenter society at large. At this time they want reassurance that society will afford them as much social security as the group has during the winter. We interpret their hunger for praise and recognition at this time as illustrating this need for assurance that they still have an established and meaningful place in society as well as in their families and places of employment. Anything that can be interpreted as a sign of rejection by outside superiors, such as delays in the arrival of relief parties, shortages in supplies, or actions of authority that interfere with established routines or disappoint seemingly trivial expectations, emphasizes the group's dependent status and may seriously threaten feelings of personal autonomy and competence as well as raise the men's reentry anxiety levels and unduly undermine their psychological well-being.

FUTURE RESEARCH

As the USARP activities have evolved over the past few years, there has been a trend for more emphasis on civilian participation in all phases of the operation and a diminishing involvement of the Navy in logistics operations. Eventually, this trend will alter the group configurations of the U.S. antarctic stations. To some extent the groups may become more homogeneous, but they may also become heterosexual. Women working in the Antarctic have recently had an opportunity to journey to the south pole as 'tourists.' In November 1973 two female technicians from the University of Oklahoma Health Sciences Center spent 2 weeks at South Pole station and will be remembered as the first women to be engaged in productive work at that location. The experiences of two women who previously wintered in the Antarctic with their husbands have been described by *Darlington and McIlvaine* [1956]. However, the problems involved in establishing a heterosexual wintering station party or an all-female group remain to receive serious consideration by behavioral scientists.

Taylor [1969*b*] appears to have been the only one to address this problem. He believes that heterosexual parties in the Antarctic will require men with different personalities from those who have wintered there in the past. Taylor's study of the behavior of women in penal institutions, a psychological environment that may have some similarities to that of the isolated station, suggests that women isolated without men do not maintain the same integrity, efficiency, and group cohesion as men do when they are isolated without women. In any event, much remains to be explored concerning the problems of men in isolation (including sexual deprivation), and the problems of women who have men in Antarctica, as well as those of women who may go to the Antarctic, are demanding attention.

As a final word we would indicate that we believe the insights gained by the study of human responses to life in Antarctica by healthy men may have direct application to the problems of men and women in urban society. Many individuals find themselves psychologically, if not physically, isolated from their neighbors for one reason or another in the mundane life of an industrial society. *Shneidman* [1967, p. 535] reminds us that most of life is so dull that there is little to be said about it and that we are not as alert or active in our environment as we would like to be. Does this apparent withdrawal from life reflect increasing stress due to restrictive, hazardous, and overcrowded conditions present in the urban scene? Studies conducted in Antarctica may give us new insight into the biosocial conditions that elicit stress responses, the time required for adaptation to stress, and some indication of the physiological as well as the psychosocial price of prolonged exposure to the confining and inhospitable aspects of the contemporary urban environment.

REFERENCES

Altman, I., An ecological approach to the functioning of socially isolated groups, in *Man in Isolation and Confinement,* edited by J. Rasmussen, pp. 241–269, Aldine, Chicago, Ill., 1973.

Altman, I., and W. W. Haythorn, Interpersonal exchange in isolation, *Sociometry, 28,* 411–426, 1965.

Altman, I., and W. W. Haythorn, The ecology of isolated groups, *Behav. Sci., 12,* 169–182, 1967*a*.

Altman, I., and W. W. Haythorn, The effects of social isolation and group composition on performance, *Hum. Relat., 20,* 313–340, 1967*b*.

Bales, R. F., *Interaction Process Analysis: A Method for the Study of Small Groups,* p. 33, Addison-Wesley, Reading, Mass., 1950.

Blackburn, A. B., J. T. Shurley, and K. Natani, Psychological adjustment at a small antarctic station: An MMPI study, in *Polar Human Biology,* edited by O. G. Edholm and E. K. E. Gunderson, pp. 369–383, Heinemann Medical Books, London, 1973.

Boag, T. J., The white man in the Arctic: A preliminary study of the problems of adjustment, *Amer. J. Psychiat., 109,* 444–449, 1952.

Bogoslovskii, M. M., Actographic study of human sleep in the Antarctic, in *Medical Research on Arctic and Antarctic Expeditions,* edited by A. L. Matusov, translated from Russian by Israel Program for Scientific Translations, *Access. TT 72-50004,* pp. 104–106, National Technical Information Service, Springfield, Va., 1973.

Bradburn, N. M., *The Structure of Psychological Well-Being,* pp. 123–146, Aldine, Chicago, Ill., 1969.

Bradburn, N. M., and D. Caplovitz, *Reports on Happiness: A*

Pilot Study of Behavior Related to Mental Health, pp. 8–135, Aldine, Chicago, Ill., 1965.

Chapple, E. D., Measuring human relations: An introduction to the study of the interaction of individuals, *Genet. Psychol. Monogr., 22,* 3–147, 1940.

Darlington, J., and J. McIlvaine, *My Antarctic Honeymoon,* pp. 1–284, Doubleday, New York, 1956.

Doll, R. E., and E. K. E. Gunderson, Occupational group as a moderator of the job satisfaction-job performance relationship, *J. Appl. Psychol., 53,* 359–361, 1969*a*.

Doll, R. E., and E. K. E. Gunderson, Hobby interest and leisure activity behavior among station members in Antarctica, *Unit Rep. 69–34,* pp. 6–17, U.S. Navy Med. Neuropsychiat. Res. Unit, San Diego, Calif., 1969*b*.

Doll, R. E., E. K. E. Gunderson, and D. H. Ryman, Relative predictability of occupational groups and performance criteria in an extreme environment, *J. Clin. Psychol., 25,* 399–402, 1969.

Eberhard, J. W., The problem of off duty time in long duration space missions, *Rep. TR-55-67-14(U),* vol. 2, pp. 9–64, Serendipity Ass., McLean, Va., 1967.

Eddowes, E. E., Survey of leisure time activity, Implications for the design of a space vehicle, *Aerosp. Med., 32,* 541–544, 1961.

Eilbert, L. E., and R. Glaser, Differences between well and poorly adjusted groups in an isolated environment, *J. Appl. Psychol., 43,* 271–274, 1959.

Festinger, L., A theory of social comparison processes, *Hum. Relat., 7,* 117–140, 1954.

Fiske, D. W., and S. R. Maddi (Eds.), A conceptual framework, in *Functions of Varied Experience,* pp. 11–56, Dorsey, Homewood, Ill., 1961.

Francis, H. S., Jr., and R. W. Mason, Antarctica–Why?, *Amer. Alpine J., 14,* 377–396, 1965.

Fraser, T. H., Leisure and recreation in long duration space missions, *Hum. Factors, 10,* 483–488, 1968.

Goffman, E., *Asylums: Essays on the Social Situation of Mental Patients and Other Inmates,* pp. 3–124, Doubleday, New York, 1961.

Gunderson, E. K. E., Emotional symptoms in extremely isolated groups, *Arch. Gen. Psychiat., 9,* 362–368, 1963.

Gunderson, E. K. E., Adaptation to extreme environments: The antarctic volunteer, *Unit Rep. 66-4,* p. 24, U.S. Navy Med. Neuropsychiat. Res. Unit, San Diego, Calif., 1966*a*.

Gunderson, E. K. E., Selection for antarctic service, *Unit Rep. 66-15,* pp. 8–11, U.S. Navy Med. Neuropsychiat. Res. Unit, San Diego, Calif., 1966*b*.

Gunderson, E. K. E., Adaptation to extreme environments: Prediction of performance, *Unit Rep. 66-17,* pp. 11–12, U.S. Navy Med. Neuropsychiat. Res. Unit, San Diego, Calif., 1966*c*.

Gunderson, E. K. E., Mental health problems in Antarctica, *Arch. Environ. Health, 17,* 558-564, 1968.

Gunderson, E. K. E. (Ed.), Psychological studies in Antarctica, in *Human Adaptability to Antarctic Conditions, Antarctic Res. Ser.,* vol. 22, this volume, AGU, Washington, D. C., 1974.

Gunderson, E. K. E., and J. L. Mahan, Cultural and psychological differences among occupational groups, *J. Psychol., 62,* 287–304, 1966.

Gunderson, E. K. E., and P. D. Nelson, Adaptation of small groups to extreme environments, *Aerosp. Med., 34,* 1111–1115, 1963.

Gunderson, E. K. E., and P. D. Nelson, Personality differences among Navy occupational groups, *Personnel Guid. J., 45,* 956–961, 1966*a*.

Gunderson, E. K. E., and P. D. Nelson, Criterion measures for extremely isolated groups, *Personnel Psychol., 19,* 67–80, 1966*b*.

Haggard, E. A., Isolation and personality, in *Personality Change,* edited by P. Worchel and D. Byrne, pp. 433–469, John Wiley, New York, 1964.

Harris, T. A., *I'm OK, You're OK: A Practical Guide to Transactional Analysis,* p. 115, Harper and Row, New York, 1967.

Haythorn, W. W., The miniworld of isolation: Laboratory studies, in *Man in Isolation and Confinement,* edited by J. Rasmussen, pp. 219–239, Aldine, Chicago, Ill., 1973.

Haythorn, W. W., and I. Altman, Personality factors in isolated environments, in *Psychological Stress,* edited by M. H. Appley and R. Trumbull, pp. 363–369, Appleton-Century-Crofts, New York, 1967.

Haythorn, W. W., I. Altman, and T. I. Myers, Emotional symptomatology and subjective stress in isolated pairs of men, *J. Exp. Res. Personality, 1,* 290–305, 1966.

Homans, G. C., *The Human Group,* pp. 453–463, Harcourt, Brace, and World, New York, 1950.

Il'in, E. A., Changes in neuropsychological condition and drug methods of its normalization during wintering at Vostok station, in *Medical Research on Arctic and Antarctic Expeditions,* edited by A. L. Matusov, translated from Russian by Israel Program for Scientific Translations, *Access. TT 72-50004,* pp. 73–79, National Technical Information Service, Springfield, Va., 1973.

Il'in, E. A., and B. B. Egorov, Extended exposure of man to sensory deprivation, *Aerosp. Med., 41,* 1022–1030, 1970.

Janis, I. L. (Ed.), Grief, in *Personality: Dynamics, Development, and Assessment,* pp. 171–198, Harcourt, Brace, and World, New York, 1969.

Jennings, H. H., *Leadership and Isolation: A Study of Personality in Interpersonal Relations,* 2nd ed., pp. 227–273, Longmans, Green, New York, 1950.

Jones, M. B., and J. E. Goodson, The effect of boredom on suggestibility, *Aerosp. Med., 30,* 716–721, 1959.

Kral, V. A., Psychiatric observations under severe chronic stress, *Amer. J. Psychiat., 108,* 185–192, 1951.

Lantis, M., Environmental stresses on human behavior: Summary and suggestions, *Arch. Environ. Health, 17,* 578–585, 1968.

Law, P., Personality problems in Antarctica, *Med. J. Aust., 47,* 273–282, 1960.

Law, P., The mental strains of antarctic solitude, *Unesco Cour., 16,* 26–30, 1963.

Lewis, H. E., and J. P. Masterton, Sleep and wakefulness in the Arctic, *Lancet, 1,* 1262–1266, 1957.

Matsuda, T., Some notes on the social life of 16 persons of the Japanese antarctic wintering party (in Japanese), *Antarctic Rec., 20,* 75–87, 1964. (Engl. Transl. available, Behav. Sci. Lab., Oklahoma City Veterans Hosp., Oklahoma City, Okla.)

McGrath, J. E., and I. Altman, *Small Group Research: A Synthesis and Critique of the Field,* pp. 49–98, Holt, Rinehart, and Winston, New York, 1966.

McGuire, F., and S. Tolchin, Group adjustment at the south pole, *Brit. J. Psychiat., 107,* 954–960, 1961.

Mechanic, D., *Students under Stress: A Study in the Social Psychology of Adaptation,* pp. 38–220, Free Press, Glenco, N. Y., 1962.

Mechanic, D., Discussion on 'Self, social environment and stress' by A. Pepitone, in *Psychological Stress,* edited by M. H. Appley and R. Trumbull, pp. 199–202, Appleton-Century-Crofts, New York, 1967.

Mullin, C. S., Some psychological aspects of isolated antarctic living, *Amer. J. Psychiat., 117,* 323–325, 1960.

Nardini, J. E., R. S. Herrmann, and J. E. Rasmussen, Navy psychiatric assessment in the Antarctic, *Amer. J. Psychiat., 119,* 97–105, 1962.

Natani, K., Observations of mood and performance in a small group of men during eight months isolated duty in Antarctica, M.Sc. thesis, Dep. of Psychol., Univ. of Okla., Norman, 1971.

Natani, K., and J. T. Shurley, Disturbed sleep and effect in an extreme environment: A syndrome suggesting depletion of CNS serotonin in men wintering at the south pole, in *Sleep: Physiology, Biochemistry, Psychology, Pharmacology, Clinical Implications,* edited by W. P. Koella and P. Levin, pp. 426–430, Karger, Basel, 1973.

Natani, K., J. T. Shurley, C. M. Pierce, and R. E. Brooks, Long-term changes in sleep patterns in men on the South Polar Plateau, *Arch. Intern. Med., 125,* 655–659, 1970.

Natani, K., J. T. Shurley, and A. T. Joern, Interpersonal relationships, job satisfaction, and subjective feelings of competence, Their influence upon adaptation to antarctic isolation, in *Polar Human Biology,* edited by O. G. Edholm and E. K. E. Gunderson, pp. 384–400, Heinemann Medical Books, London, 1973.

Nelson, P. D., Human adaptation to antarctic station life, Medicine and Public Health in the Arctic and Antarctic, *Public Health Pap. 18,* pp. 138–145, World Health Organ., Geneva, Switzerland, 1963.

Nelson, P. D., Structural change in small isolated groups, *Unit Rep. 64-24,* 12 pp., U.S. Navy Med. Neuropsychiat. Res. Unit, San Diego, Calif., 1964.

Nelson, P. D., The indirect observation of groups under confinement and/or isolation, in *Man in Isolation and Confinement,* edited by J. Rasmussen, pp. 167–193, Aldine, Chicago, Ill., 1973.

Oken, D., The role of defense in psychological stress, in *Physiological Correlates of Psychological Disorder,* edited by R. Roessler and N. S. Greenberg, pp. 193–210, University of Wisconsin Press, Madison, 1962.

Palmai, G., Psychological aspects of transient populations in Antarctica, Medicine and Public Health in the Arctic and Antarctic, *Public Health Pap. 18,* pp. 146–158, World Health Organ., Geneva, Switzerland, 1963*a*.

Palmai, G., Psychological observations on an isolated group in Antarctica, *Brit. J. Psychiat., 109,* 364–370, 1963*b*.

Pepitone, A., The role of danger in affiliation and attraction, *Acta Psychol., 18,* 1–10, 1961.

Pepitone, A., *Attraction and Hostility: An Experimental Analysis of Interpersonal and Self-Evaluation,* pp. 5–49, Atherton, New York, 1964.

Pepitone, A., Self, social environment and stress, in *Psychological Stress,* edited by M. H. Appley and R. Trumbull, pp. 182–207, Appleton-Century-Crofts, New York, 1967.

Pierce, C. M., A philosophic statement: Relevance of antarctic biomedical research to society in the 70's, in *Biomedical and Behavioral Science Research in Antarctica,* pp. 7–14, National Academy of Sciences, Washington, D. C., 1971.

Popkin, M. K., V. Stillner, L. W. Osborn, C. M. Pierce, and J. T. Shurley, Novel behaviors in an extreme environment, *Amer. J. Psychiat., 131,* 651–654, 1974.

Priestley, R., Inexpressible Island, *Nutr. Today, 4,* 18–27, 1969.

Radloff, R. W., Naturalistic observations of isolated experimental groups in field settings, in *Man in Isolation and Confinement,* edited by J. Rasmussen, pp. 195–216, Aldine, Chicago, Ill., 1973.

Radloff, R. W., and R. Helmreich, *Groups under Stress: Psychological Research in SEALAB 2,* pp. 97–155, 177–210, Appleton-Century-Crofts, New York, 1968.

Rioch, M. J., Group relations: Rational and techniques, *Int. J. Group Psychother., 20,* 340–355, 1970.

Rohrer, J. H., Interpersonal relationships in isolated small groups, in *Psychophysiological Aspects of Space Flight,* edited by B. E. Flaherty, pp. 263–271, Columbia University Press, New York, 1961.

Ruff, G. E., Psychological effects of space flight, *Aerosp. Med., 32,* 639–642, 1961.

Ruff, G. E., E. Z. Levy, and V. H. Thaler, Studies of isolation and confinement, *Aerosp. Med., 30,* 599–604, 1959.

Schachter, S., *The Psychology of Affiliation: Experimental Studies of the Sources of Gregariousness,* pp. 2–133, Stanford University Press, Stanford, Calif., 1959.

Sells, S. B., and E. K. E. Gunderson, A social systems approach to long-duration missions, in *Human Factors in Long-Duration Space Flight,* edited by D. B. Lindsley, pp. 179–208, National Academy of Sciences, Washington, D. C., 1972.

Shapiro, D., and G. E. Schwartz, Psychophysiological contributions to social psychology, *Ann. Rev. Psychol., 21,* 87–111, 1970.

Shneidman, E. S. (Ed.), Sleep and self-destruction: A phenomenological approach, in *Essays in Self-Destruction,* pp. 510–539, Science House, New York, 1967.

Shurley, J. T., Man on the South Polar Plateau, *Arch. Intern. Med., 125,* 625–629, 1970.

Shurley, J. T., New directions in antarctic biomedical research, in *Research in the Antarctic,* edited by L. O. Quam, pp. 277–287, AAAS, Washington, D. C., 1971.

Shurley, J. T., C. M. Pierce, K. Natani, and R. E. Brooks, Sleep and activity patterns at South Pole station, *Arch. Gen. Psychiat., 22,* 385–389, 1970.

Siple, P. A., *90° South: The Story of the American South Pole Conquest,* pp. 309–312, Putnam, New York, 1959.

Smith, W. M., Observations over the lifetime of a small isolated group: Structure, danger, boredom, and vision, *Psychol. Rep., 19, Monogr. Suppl. 3-V19,* 475–514, 1966.

Strickland, D. A., Preliminary thoughts on the politics of closed environments, *Intern. Work. Pap. 16,* pp. 1–31, Space Sci. Lab., Univ. of Calif., Berkeley, 1964.

Strickland, D. A., A note on proto-political behaviors, *Intern. Work. Pap. 24,* pp. 3–13, Space Sci. Lab., Univ. of Calif., Berkeley, 1965.

Suedfeld, P., Attitude manipulation in restricted environments, 1, Conceptual structure and response to propaganda, *J. Abnorm. Soc. Psychol., 68,* 242–247, 1964.

Suedfeld, P., Changes in intellectual performance and in susceptibility to influence, in *Sensory Deprivation: Fifteen Years of Research,* edited by J. P. Zubek, pp. 126–166, Appleton-Century-Crofts, New York, 1969.

Taylor, A. J. W., Professional isolates in New Zealand's antarctic research programme, *Int. Rev. Appl. Psychol., 18,* 135–138, 1969*a*.

Taylor, A. J. W., Ability, stability, and social adjustment among Scott base personnel, Antarctica, *Occup. Psychol., 43,* 81–93, 1969*b*.

Taylor, A. J. W., and J. T. Shurley, Some antarctic troglodytes, *Int. Rev. Appl. Psychol., 20,* 143–148, 1971.

Taylor, D. A., L. Wheeler, and I. Altman, Stress relations in socially isolated groups, *J. Personality Soc. Psychol., 9,* 369–376, 1968.

Vallacher, R. R., G. E. Seymour, and E. K. E. Gunderson, The relationship between cohesiveness and effectiveness in small isolated groups: A field study, *Unit Rep. 74-50,* pp. 11–14, U.S. Navy Neuropsychiat. Res. Unit, San Diego, Calif., 1974.

Walker, H. M., and J. Lev, *Statistical Inference,* pp. 151–154, Henry Holt, New York, 1953.

Walters, R. H., W. E. Marshall, and R. J. Shooter, Anxiety, isolation and susceptibility to social influence, *J. Personality, 28,* 518–529, 1960.

Weller, L., The effects of anxiety on cohesiveness and rejection, *Hum. Relat., 16,* 189–197, 1963.

Wheaton, J. L., Fact and fancy in sensory deprivation studies, *Aeromed. Rev. 5-59,* Air Univ. School of Aviat. Med., U.S. Air Force, Brooks Air Force Base, Tex., 1959.

White, R. W., Motivation reconsidered: The concept of competence, in *Functions of Varied Experience,* edited by D. W. Fiske and S. R. Maddi, pp. 278–325, Dorsey, Homewood, Ill., 1961.

Wilkins, H., Discussion on 'The white man in the Arctic: A preliminary study of the problems of adjustment' by T. J. Boag, *Amer. J. Psychiat., 109,* 444–449, 1952.

Wilkins, W. W., Isolation research; The methodological context, in *Man in Isolation and Confinement,* edited by J. Rasmussen, p. 315, Aldine, Chicago, Ill., 1973.

Wilson, O., Human adaptation to life in Antarctica, in *Biogeography and Ecology in Antarctica, Monogr. Biol.,* vol. 15, edited by J. van Meighem, P. van Oye, and J. Schell, pp. 690–752, W. Junk, The Hague, Netherlands, 1965.

PSYCHOLOGICAL STUDIES IN ANTARCTICA

E. K. ERIC GUNDERSON

Epidemiology and Operational Psychiatry Division, U.S. Navy Medical Neuropsychiatric Research Unit San Diego, California 92152

Abstract. Clinical evaluations by psychologists and psychiatrists, biographical information, and attitude and personality tests have been used in selection studies of antarctic personnel since 1963. A series of studies using supervisor ratings and peer nominations as performance criteria resulted in a general concept of effective individual performance that included three essential behavioral components: emotional stability, task motivation, and social compatibility. The contributions of the various sources of screening information to the prediction of individual and group adjustment during the antarctic winter were determined, and a high degree of specificity in the relevance of screening predictors for the various behavioral criteria and occupational subgroups was found. Navy men showed significant deterioration in morale or satisfaction during the winter months, whereas civilians showed little or no change. Thus occupational role was an important determinant of job satisfaction during long-term isolation and confinement.

Psychological studies were initiated at U.S. stations by the U.S. Navy Bureau of Medicine and Surgery and the U.S. Office of Naval Research during the International Geophysical Year (IGY) in 1957–1958. Attitude and symptom questionnaires, supervisor ratings, and sociometric tests were administered to several wintering groups during 1957 and 1958. These tests were designed by Herbert Zimmer of the Georgetown University Medical School, Washington, D. C.; the data were transferred to the U.S. Navy Medical Neuropsychiatric Research Unit, San Diego, California, for analysis in 1961.

Mullin and Connery [1959], a psychiatrist-psychologist investigating team, interviewed and tested members of Ellsworth station wintering-over groups in 1957–1958. *Mullin* [1960] and *Rohrer* [1961] visited several antarctic stations in 1958–1959 and gathered interview observations concerning environmental stresses and group and individual adjustment problems.

Smith [1961] and *Smith and Jones* [1962] studied methods of selecting scientific personnel during the IGY. *Smith* [1966] also studied group structure and group behavior in a traverse party that faced extreme danger as well as monotony and boredom. This work was supported by the Office of Antarctic Programs, National Science Foundation.

McGuire and Tolchin [1961] analyzed test responses, diary observations, and medical records of the South Pole group wintering in 1959 and evaluated individual and group adjustment at that station.

Nardini et al. [1962] described the psychiatric assessment program for selection of U.S. personnel during and immediately after the IGY and reported that psychiatric evaluations had been relatively successful in predicting antarctic performance as measured by supervisor ratings.

Palmai [1963*a, b*] reported on extensive test and observation data gathered in a group of 14 men who wintered at Macquarie Island. This thorough and informative study was not repeated, but results were nevertheless suggestive. Palmai noted a decline in morale, an increase in group conflict, and an increase in dispensary visits in the third quarter of the year. He called attention to some of the difficulties of the participant observer's role in an isolated antarctic group.

A new program of psychological studies, sponsored by the Bureau of Medicine and Surgery and directed by the author, was instituted at the Navy Medical Neuropsychiatric Research Unit in 1961. In the beginning of this program, test and rating data collected during the IGY and immediately thereafter were thoroughly analyzed to develop hypotheses and a methodological framework for a long-range plan of psychological investigations. The observations and interview data gathered by psychologist-psychiatrist debriefing teams during the IGY provided valuable information concerning individual and group reactions to antarctic isolation; questionnaire and performance rating data gathered during the same period, although they were incomplete, also aided in the development of a systematic research program.

DESCRIPTION OF U.S. STATIONS

The major studies reviewed involve wintering-over groups during 1963–1969 at six U.S. stations: South

Pole, Byrd, Eights, Hallett, Palmer, and Plateau. Important features of these six stations are described in some detail.

The Amundsen-Scott South Pole station, located at the geographical south pole, is approximately 2800 meters above sea level. At this elevation the barometric pressure ranges from 485 to 525 mm Hg, equivalent to the pressure at 3350 meters of altitude elsewhere. The air is extremely dry. The mean annual temperature at South Pole is -57°F, and the extremely low temperature of -121°F has been recorded. At the south pole there is an annual cycle of 6 months of light and 6 months of darkness. The station, supplied entirely by air, is completely isolated and inaccessible to the outside world from February until November except for intermittent radio communication. Buildings at South Pole, although they were built on the surface, are now in tunnels under the ice because of the accumulation of drifting snow. Only a few structures containing scientific equipment are on the surface. The crushing weight of accumulating snow and the constant movement of the ice in spite of constant reinforcement of wooden and metal beams supporting the tunnel roof threaten the station with collapse. Structurally, South Pole station is in poor condition, and a new station is being built and should be completed in 1975. The new station will include facilities for biomedical research.

Byrd station is on the interior ice sheet and at an elevation of about 1525 meters above sea level. The prefabricated, insulated buildings were placed in long trenches 12 meters deep roofed over with steel arches and packed snow. The slow movement of the ice sheet bends the arches and pushes the tunnel walls together, a constant effort thus being required to counteract the damaging effects.

Two small inland stations, Eights and Plateau, were constructed by air-hauling prefabricated vans to the sites. At Eights, located at an elevation of 460 meters near the base of the Antarctic Peninsula, eight vans were arranged in two parallel rows, and flooring and roofing were placed between the rows, a large nine-room building being created. This station opened in 1963 and closed at the end of the 1965 season. Plateau station was constructed in a similar manner with four vans in January 1966. This station was located 970 km from South Pole, near the center of the South Polar Plateau, at an elevation of 3660 meters. This site, together with the Russian Vostok station, represents the most hostile environment ever endured for long periods by man. The record low temperature at Plateau was -121°F, and the lowest temperature ever recorded, -127°F, was registered at Vostok.

Hallett and Palmer stations are both located on the coast and have milder climates than the interior stations. Hallett was one of the original IGY stations and was operated jointly by the United States and New Zealand from 1957 through 1964. By 1965, Hallett had nine buildings for scientific use and 13 for general use, there being approximately 418 m^2 for living areas. Since 1965 the station has been occupied during summer only. The station is located 610 km from McMurdo station at sea level and has a mean annual temperature of 4.2°F.

Palmer station is located on Anvers Island along the west coast of the Antarctic Peninsula. The British maintained a station there from 1955 to 1958, and the United States added new buildings and occupied the site in 1965. In 1967–1968 a new two-story building was constructed about 1600 meters from the old station; the living area was increased from 172 m^2 at the old station to more than 470 m^2 at the new station. Palmer station has a relatively mild climate with a mean annual temperature of 26°F.

Table 1 indicates the numbers and types of personnel wintering at U.S. stations during 1963–1969.

EARLY STUDIES

Three major types of screening information were used for selection studies from 1963 to the present: (1) clinical evaluations by psychologists and psychiatrists, (2) biographical information, and (3) attitude and personality tests. Two of the personality questionnaires, the friend description and Firo-B inventories, were originally included in the test battery to explore certain hypotheses pertaining to interpersonal compatibility in wintering-over groups.

The early development and evaluation of clinical rating procedures will be described first; the results of recent large-scale studies of the predictive validities of clinical ratings will be presented in the section on prediction of individual adjustment.

Three areas of research on clinical ratings were of interest: interrater reliability, similarities and differences in clinicians' evaluative frames of reference, and prediction of antarctic adjustment. Interrater reliability was studied in a sample of 719 Navy volunteers for antarctic service [*Gunderson,* 1965*a*]. Each candidate was independently examined and rated on a series of personality traits and adjustment predictions by a psychiatrist and a clinical psychologist. Reliabilities were compared under four different assessment conditions: (1) similar examining methods (both clinicians used interviews), (2) dissimilar examining methods (psychiatrists used interviews, and psychologists administered Rorschach tests),

TABLE 1. U.S. Wintering Party Composition by Occupational Subgroup, Station, and Year

Year	Station	Navy Officers	Navy Enlisted Personnel		Civilian Personnel		Total
			Construction-Maintenance	Administrative-Technical	Scientists	Meteorologists-Technicians	
1963	Byrd	2	14	5	6	10	37
	Hallett	1	3	5	1	5	15
	South Pole	1	7	5	4	5	22
	Eights	...	2	4	3	2	11
1964	Byrd	1	10	6	4	6	27 (1)
	Hallett	1	4	6	...	...	11 (2)
	South Pole	1	6	6	5	4	22
	Eights	...	2	4	3	1	10
1965	Byrd	2	11	6	5	4	28
	South Pole	1	7	6	4	3	21
	Eights	...	2	4	2	3	11
	Palmer	...	1	3	5	...	9
1966	Byrd	2	11	5	5	7	30
	South Pole	1	6	5	4	2	18
	Plateau	1	1	2	2	2	8
	Palmer	...	1	3	3	1	8
1967	Byrd	1	11	5	6	4	27 (1)
	South Pole	1	7	5	3	4	20 (1)
	Plateau	1	2	1	2	...	6 (2)
	Palmer	...	1	3	4	1	9
1968	Byrd	2	8	6	5	5	26
	South Pole	2	5	5	4	3	19
	Plateau	1	1	2	2	2	8
	Palmer	1	2	3	2	1	9
1969	Byrd	2	8	6	4	3	23
	South Pole	2	6	6	3	3	20
	Palmer	2	3	3	2	3	13

Navy officers include medical officers and engineering officers; construction-maintenance personnel include carpenters, mechanics, electricians, plumbers, and tractor operators; administrative-technical personnel include electronics technicians, radio operators, hospital corpsmen, storekeepers, and cooks; scientists include college graduates assigned to programs in ionospheric physics, aurora, radio science, cosmic rays, seismology, geomagnetics, biology, etc.; meteorologists-technicians generally include non–college graduates assigned to meteorology and electronics (a few were college graduates).

Numbers in parentheses indicate scientists from other countries.

(3) clinicians experienced in working together, and (4) clinicians not experienced in working together. Agreement was better when similar clinical methods (interviews) were used, but no difference in agreement was apparent between pairs of interviewers experienced in working together and those not experienced in working together. Better reliabilities were achieved on global rating items than on specific traits, and the highest agreement was obtained on ratings of overall effectiveness (r = 0.47). A previous study [*Gunderson and Nelson,* 1964] had shown that another sample of Deep Freeze examiners was in high agreement in global assessments of potential effectiveness.

In spite of substantial agreement on global adjustment ratings it was apparent that antarctic examiners often differed in their impressions and descriptions of personality. A study was conducted of the variability in rating patterns among 27 psychologists and psychiatrists who participated in the screening program in 1963–1964. Ratings of personality traits and adjustment predictions were factor-analyzed for each of the 27 clinicians individually [*Gunderson and Kapfer,* 1966*a*].

Factor patterns and trait meanings were found to be generally consistent among the psychologists and psychiatrists on a rating schedule that contained only common trait adjectives. Factor structures were somewhat less consistent on a schedule including specialized Rorschach terminology and psychodynamic concepts.

Essentially the same three major factors emerged for 10 psychologists and 16 psychiatrists on the schedule containing common trait adjectives: a general evaluative factor, an aggression factor, and an efficiency factor. The general evaluative factor was clearly represented in the factor patterns of all individual clinicians, the efficiency factor was present for all but two clinicians, and the aggression factor was discernible in the factor patterns of all but four clinicians.

The study demonstrates the feasibility of combining trait ratings into stable clusters, or scales, having similar meanings for the majority of examiners in the antarctic screening program.

Factors affecting interrater reliabilities for clinical ratings were analyzed in one study [*Gunderson,* 1965*b*]. Fifteen of the 35 teams studied achieved reliability coefficients for the overall effectiveness item exceeding 0.50; there were wide differences in agreement among teams, however. The effects of variations in professional experience and idiosyncrasies of clinicians' rating patterns on interrater reliabilities were examined. Similarity of professional experience, favorability of ratings, deviations from modal rating profiles, and agreement of the importance of the various traits for the antarctic assignment were found to be related to interrater agreement. Much more needs to be done to understand how trait meanings and the processes of cue selection, evaluation, and combination affect clinician agreement.

Early validity studies of psychiatric evaluations were discussed by *Gunderson and Kapfer* [1966*b*]. During the IGY, clinical ratings correlated substantially with supervisors' ratings of adjustment in Antarctica, but, in later expeditions, useful validity was more difficult to achieve. The relative contribution of clinical ratings to the prediction of antarctic adjustment in current studies is considered in the section on prediction of individual adjustment.

BIOGRAPHICAL INFORMATION

Life history and occupational experience data are commonly used in the selection of persons for specialized jobs in industry, government, and the military. In antarctic research, biographical information has been used for three purposes: (1) to predict individual adjustment or performance; (2) to differentiate occupational subgroups in terms of educational, familial, social, and cultural background characteristics; and (3) to predict group cohesiveness or compatibility on the basis of diversity or similarity in biographical characteristics.

In two studies of Navy personnel from large stations (Little America and McMurdo) during 1957–1958 and 1960–1961, a number of biographical items were found to be related to supervisor ratings or peer evaluations or both. Age, job experience, and pay grade (rank) were positively related to performance evaluations in both samples. In later studies of small-station personnel (1960–1964), the oldest, most experienced individuals were given the highest performance ratings, but the relationship between age and performance was not linear. The youngest men also tended to receive high ratings, largely based on social popularity, however. Men with delinquency-truancy records received poor performance ratings at both large stations and small stations. The most striking finding in this series of studies was the difference between the significance of recreational interests and activities at large stations and that at small stations. At large stations, where recreational and social activities were more varied and plentiful, expressed interest in many avocational activities was positively correlated with adjustment, whereas at small stations, where opportunities for recreational activities were very limited, expressed preference for many hobbies and leisure activities tended to be negatively correlated with adjustment [*Gunderson and Nelson,* 1965*a*]. Many life history and demographic variables were found to be unrelated to antarctic performance in these early studies, and biographical information seemed to contribute little to personnel assessment for Navy personnel in this setting. Later studies using larger samples indicated that biographical attributes were effective as predictors of performance, particularly for the civilian subgroup. These results are considered in the section on prediction of individual adjustment.

ATTITUDE AND PERSONALITY TESTS

An attitude questionnaire describing motivation for the antarctic assignment, confidence in expedition leadership, satisfaction with Navy duty, and reactions to cold weather, food, and clothing was administered in the screening programs for the 1958 and 1961 expeditions. Analysis of responses in relation to wintering-over adjustment ratings resulted in a major revision of the questionnaire in 1962 and an expansion of the item content in 1963 and 1964. Since 1964 an attitude questionnaire called the opinion survey has been administered routinely in the screening program. Five scales derived from this inventory were carried over from the original attitude questionnaire, and these measure attitudes toward the antarctic assignment. They are called motivation, job importance, boredom, confidence in organization, and confidence in medical care. The factor analytic technique was employed to construct these and other test scales designed especially for the antarctic screening program. Four scales measure common psychological needs: achievement, autonomy, nurture, and orderliness. The contents of these scales are generally similar to those of the corresponding Edwards personal preference schedule scales [*Edwards,* 1959]. The Edwards inventory was administered in the screening program for the 1958 expedition, and item analysis indicated that the four scales listed above are most relevant to antarctic adjustment.

Four scales represent personality self-descriptions based on clusters of intercorrelated traits: decisiveness (decisive, obedient, handy, alert, orderly, and self-improving), excitability (excitable, argumentative, hostile, suspicious, and impulsive), bluntness (blunt, stubborn, rough in manner, and hard), and absentmindedness (absentminded, awkward, and slow).

Seven scales were derived from an inventory consisting of 50 trait adjectives rated by subjects on 8-point scales in terms of how much the individual would like these characteristics in a close friend. The scales are labeled efficiency, sympathy, conservatism, optimism, dignity, cynicism, and humor. This instrument was developed to study the predictability of friend choices in antarctic groups. It seems plausible that at least some individuals could describe the personality characteristics of persons whom they would find likable.

The only published test used consistently in antarctic research has been the Firo-B questionnaire. This inventory purports to measure two aspects (expressing and wanting) of three fundamental interpersonal needs [*Schutz,* 1958]. The six scales may be characterized as follows: expressed inclusion (participating in group activities), wanted inclusion (wanting to be included in group activities), expressed control (controlling others, exercising dominance or leadership), wanted control (wanting direction or regulation from others), expressed affection (being affectionate with others), and wanted affection (wanting affection from others). The Firo-B was first given to wintering-over personnel during the IGY. Preliminary work indicated that the scales differentiate sharply between occupational subgroups and that certain scales correlate with social adjustment in antarctic groups. The test has been included in routine psychological screening since 1963.

The Allport-Vernon-Lindzey study of values [*Allport et al.,* 1951] and the survey of interpersonal values [*Gordon,* 1960], which purport to measure basic motivational patterns and value orientations, were administered as part of antarctic screening for 1962 and 1963. These measures were used in a study of cultural and personality differences among antarctic occupational subgroups, and certain parts of these inventories were revised and incorporated into the opinion survey for further study.

DIFFERENCES AMONG ANTARCTIC OCCUPATIONAL SUBGROUPS

Analysis of IGY biographical data had indicated that the Navy and the civilian subgroups differed on many background characteristics, of which age, education, and occupational experience are the most notable. After the IGY the age difference between the two subgroups was reduced, since the civilian participants tended to be younger (graduate students rather than senior scientists), whereas the Navy participants were generally older and more experienced than they had been previously. Because social background and psychological differences could confound the prediction of individual performance and possibly affect group compatibility and effectiveness, a detailed analysis of cultural and psychological differences among antarctic occupations was undertaken [*Gunderson and Mahan,* 1966]. In the first phase of the study, 1405 Navy and civilian volunteers were grouped into nine occupational categories and were compared on a number of social background characteristics contained in a biographical questionnaire. In the second phase a new sample of 717 antarctic volunteers was divided into the same occupational categories, and differences among subgroups were examined for the 47 test scales and personality ratings described in the previous section.

Navy enlisted occupations were grouped into the six categories shown below. The principal occupational specialties making up each category are indicated.

Personnel	Occupations
Administrative	Storekeeper, hospital corpsman
Technical	Radioman, electronics technician
Mechanical	Construction mechanic, engineman (diesel mechanic)
Construction	Builder, utility man (plumber)
Electrical	Construction electrician, electrician's mate
Cooks	Commissary man

Three additional occupational subgroups were formed: one consisted entirely of naval officer volunteers, another consisted entirely of scientists, and the third consisted entirely of meteorologists and civilian technicians.

The nine antarctic occupational subgroups varied widely in social and cultural background characteristics. The subgroups were essentially dichotomized on college attendance, almost all the scientists and officers, as opposed to relatively few of the Navy enlisted men, having attended college. The various occupations differed sharply on father's occupation, and father's occupation was highly related to the individual's own occupation. A large proportion of the fathers of scientists and officers were in white-collar jobs, whereas the fathers of Navy enlisted men were predominantly in blue-collar trades or farming. Parents' education also varied considerably among occupational subgroups; these differences were consistent with differences in father's education. Scientists and officers tended to reside in urban areas and in the northeastern United States more frequently than other subgroups. Navy men more often resided in rural areas or in the southern states. Scientists and officers generally reported interests in several hobbies, whereas Navy enlisted men indicated few avocational activities.

The scientist and officer subgroups did not differ significantly from each other on any of the cultural attributes studied. Civilian technicians (mostly meteorologists) differed consistently from scientists and officers, but Navy enlisted men differed from those subgroups much more.

Differences in psychological test results were examined for five occupational subgroups, i.e., scientists, officers, technical personnel, mechanical personnel, and cooks, representing well the range and nature of the differences among the nine original subgroups. Large differences were present among occupational subgroups on most of the personality scales. On many of the scales, scientists and officers tended to score at one extreme, and cooks and mechanics tended to score at the other. The study clearly demonstrates that the various occupational roles are associated with specific patterns of cultural background characteristics, value systems, and psychological needs. The degree of diversity found in social backgrounds and value orientations might be expected to affect group harmony and cooperation and certainly would confound the prediction of individual behavior unless predictive studies considered the various occupational subgroups separately.

DEVELOPMENT OF INDIVIDUAL PERFORMANCE CRITERIA

The underlying assumptions and concepts guiding the construction of performance measures for antarctic stations were largely derived from a logical analysis of expedition goals and examination of preliminary data gathered during the IGY. It seemed apparent that willingness to work hard, often under extremely adverse conditions, was required if program objectives were to be met or, perhaps, if survival was to be assured. Task motivation, then, appeared to be an essential component of general effectiveness, and the IGY data, although they were incomplete, tended to confirm the importance of industriousness in overall performance evaluations.

It was soon recognized that measuring task performance in this setting would be difficult. Each of the many occupational specialties represented at the small stations had quite different tasks to perform. Furthermore, it was impossible to know in advance how much work would be required in a given area or what disruptions might occur owing to factors beyond the individual's control. One thing appeared certain: the only available source of information about an individual's work behavior was the station membership itself. It was decided that measuring work performance could best be achieved in this type of intimate, closed community by seeking a consensus of evaluations by leaders and group members.

A second area of behavior evaluation that assumed major importance in our early thinking was emotional health. The importance attributed to possible emotional problems arose largely from concern with the deleterious effects of long-term isolation and confinement and from the knowledge that the specialized skills of each group member were largely irreplaceable. During the IGY the Navy Bureau of Medicine and Surgery was assigned the responsibility for physical and psychiatric screening of all antarctic personnel. Psychiatric evaluations were intended only to identify and disqualify potential psychotic or seriously disturbed individuals. Within a population of healthy volunteers it was considered difficult if not impossible for examiners to differentiate individuals in terms of probable future adjustment in the Antarctic without having much more relevant information about the environmental and social conditions. Fortunately, cases of psychosis or severe neurosis have been extremely rare at antarctic stations, so that severe psychopathology has not provided a meaningful criterion for differentiating individuals on emotional behavior. Minor emotional disturbances are quite prevalent, however, as is indicated by self-report questionnaires and symptom checklists filled out by medical officers [*Gunderson,* 1963]. In early studies, station members seemed capable of observing and reporting emotional distress in their associates, and again there was no practical way for an observer outside the station to monitor an individual's emotional behavior.

The importance of a third major aspect of adjustment, social compatibility or likability, was not clearly recognized in the beginning phases of the research program. Sociometric data gathered during the IGY indicated that popularity or likability correlated substantially with overall performance. Psychiatric debriefings and station leaders' observations also emphasized the importance of social relationships in overall adjustment.

Thus the earlier studies resulted in a general concept of effective individual performance that included three essential components: emotional stability, task motivation, and social compatibility. Factor analysis of supervisors' ratings on 11 behavior traits and peer nominations on the items 'best friend' or 'easy to get along with' resulted in a rotated factor structure in which three pairs of items appeared as salient dimensions: (1) emotional control and acceptance of authority, (2) industriousness and achievement motivation, and (3) likability (from supervisor ratings) and friendship and compatibility (from peer nominations). These three factors tended to have the same multiple correlation relationship (r = 0.89) with a general adjustment criterion 'would select to serve with again in the Antarctic' for various subgroups of personnel, and the addition of other characteristics did not improve the multiple correlation. These initial efforts to measure separate aspects, or components, of desired performance appeared to be at least partially successful in identifying a parsimonious, general set of behavior factors.

The next phase of the research involved collecting new criterion data over several years by means of rating and sociometric instruments. Two station leaders, the officer in charge and the scientific leader, independently filled out 10 behavior trait ratings on 8-point scales. Performance evaluations were made twice during the winter period, in March and September. A sociometric form was administered to all station members in September. Three of the sociometric items were used in deriving the three behavior factor scores. The supervisor trait items and sociometric items contributing to each of the behavior factor scores are shown in Table 2.

Group members at each station were ranked and assigned T scores (the mean is 50, and the standard deviation is 10) in terms of average supervisor ratings and number of peer nominations in each of the behavior areas. The T scores were used as controls for station size and differences in supervisors' rating distributions. Measures of the three trait areas by the two methods (supervisor ratings and peer nominations) were intercorrelated in several samples. The results are summarized in Table 3. Intercorrelations of trait areas within these data can be interpreted meaningfully in terms of the *Campbell and Fiske* [1959] model for the evaluation of convergent and discriminant validities. This method of analysis is appropriate when more than one trait area is measured by more than one method. Convergent validity is demonstrated when different methods of measuring the same trait areas significantly agree. Discriminant validity requires that correlations between measures of the same trait areas by different methods exceed correlations between different trait areas (1) across different methods and (2) within the same method. The second requirement is a very rigorous condition that has seldom been met in studies of behavior traits.

Convergent validities are in italics in the diagonals of the cross-method correlations in Table 3. All coefficients are highly significant ($p < 0.001$) and provide substantial evidence for the presence of convergent validity. Discriminant validity was determined by averaging all correlations of each trait area with other trait areas both across different methods and within the same method. Comparisons of average validities for each trait area and

TABLE 2. Supervisor and Peer Items Contributing to Behavior Criteria

Behavior Factor	Supervisor Items	Peer Items
Emotion	Emotionally stable, accepting of authority	Calm and even tempered
Task	Industrious, motivated, proficient	Industrious and hard-working
Social	Likable, cheerful, considerate of others	Friendly and popular

TABLE 3. Intercorrelations for Different Methods of Measuring Performance

	Supervisor			Peer		
	Emotion	Task	Social	Emotion	Task	Social
			Study 1			
Supervisor (128)						
Emotion						
Task	0.43					
Social	0.65	0.41				
Peer (139)						
Emotion	*0.54*	0.32	0.41			
Task	0.37	*0.54*	0.31	0.58		
Social	0.38	0.37	*0.49*	0.61	0.56	
			Study 2			
Supervisor (110)						
Emotion						
Task	0.40					
Social	0.63	0.41				
Peer (105)						
Emotion	*0.62*	0.36	0.51			
Task	0.41	*0.55*	0.28	0.48		
Social	0.48	0.25	*0.54*	0.53	0.56	
			Study 3			
Supervisor (100)						
Emotion						
Task	0.46					
Social	0.50	0.45				
Peer (109)						
Emotion	*0.65*	0.34	0.37			
Task	0.38	*0.67*	0.35	0.49		
Social	0.35	0.20	*0.47*	0.42		

Numbers in parentheses indicate the number of ratees. Numbers in italics indicate convergent validities.

average correlations with other trait areas over all samples are shown in Table 4. Discriminant validities are indicated by the fact that average validities in each case exceed average correlations with other trait areas. Further details concerning these relationships are available in two reports [*Gunderson and Nelson*, 1966; *Gunderson and Ryman*, 1971].

The overall results are most striking because of the stability or consistency in relationships from one sample to another. The development of these criteria represents a major step toward developing useful prediction of performance in future antarctic groups. A problem common to many studies in the trait domain is reflected in the relatively high correlation between the emotional behavior areas and the social behavior areas. The fact that emotional behavior and social behavior are indeed closely associated makes it difficult to achieve sharp delineation in the measurement of these concepts.

TABLE 4. Discriminant Validities of Criterion Components

	Mean Validities*	Mean Correlations with Other Areas†
Emotion	60	47
Task	59	46
Social	50	46

*Values are averaged over all samples.

†Correlations with other trait areas are averaged both across methods and within methods.

The behavior factors derived in this study appear to be equivalent to those reported by other investigators [e.g., *Borgatta*, 1964; *Owens*, 1966].

PREDICTION OF INDIVIDUAL ADJUSTMENT

With the development of a stable system of performance evaluations, predictive studies using a variety of personnel assessment methods became possible. Prediction of personnel performance in an industrial or military setting tends to be difficult even within stable and well-defined organizations. In the Antarctic we are confronted with the fact that each year new subgroups or organizations are formed that differ in some respects from previous subgroups. For example, differences in leadership, project goals, or group size or composition might change the types of demands made on individual members. In any case the need for valid selection devices remains, and the development of useful prediction equations for various job roles in this extreme situation is of urgent concern. The need to consider and integrate many kinds of information about an individual is apparent, as is the need to define as clearly as is possible the specific aspect of behavior or adjustment to be predicted.

Many studies have demonstrated that the validities of psychological predictors tend to vary over types of adjustment criteria and specific subgroups to be predicted. To obtain an overview of the antarctic selection problem, a study was conducted to evaluate the relative predictability of three antarctic occupational subgroups (Navy Seabees, technical-administrative personnel, and scientists) on five performance measures (emotion, task, social, leader, and overall) by using a variety of predictor sources [*Doll et al.*, 1969]. The findings indicate a high degree of specificity in the relevance of various predictor sources for the different criteria and occupations. In other words, biographical and personality variables that predicted social adjustment for one subgroup did not predict emotional adjustment for the same subgroup or social adjustment for another subgroup.

A more thorough and extensive analysis of a similar type was recently undertaken with larger samples of wintering-over personnel. The results are summarized in Table 5, which shows clearly the specificity described earlier. Certain sources contribute many more significant items than others for predicting adjustment in a particular subgroup. For example, opinion survey items proved much more relevant for predicting adjustment among Navy Seabees than among civilian scientists, whereas for the Firo-B test items the opposite was true. Clinical evaluations were highly relevant for only one criterion and one subgroup, leadership ability in the Seabee subgroup. It can be seen that biographical information (including hobby interests) contributed substantially to prediction in all three subgroups. The friend description items, designed or intended to predict not individual performance but, rather, dyadic friendship choices, nevertheless made minor contributions, particularly for the scientist subgroup.

The row totals in Table 5 indicate the relative predictability of the various subgroup-criteria combinations. Social adjustment among the scientists and leadership ability among the Seabees were most predictable by this indicator, and the task criterion was least predictable in both the Navy subgroups.

CHANGES IN ATTITUDES AND EMOTIONS

The individual's motivation for and satisfaction with his antarctic assignment form another important area of research. Task motivation as measured by supervisors' ratings and peer nominations is one of the principal components of antarctic performance, but the individual's own attitudes toward his job and life situation also provide useful indications of adaptation.

One objective of this research was to evaluate the ability of antarctic groups to maintain effective work motivation and positive social attitudes during winter isolation. A series of studies was conducted to identify a few attitude dimensions representing salient aspects of individual and group functioning in antarctic groups. Factor analysis was used to reduce a large number of heterogeneous test items to a small number of major scales, and this process was repeated a number of times. Two attitude inventories used in IGY expeditions, the attitude study and the group behavior description, were the source of 10 homogeneous item clusters used to measure attitude changes during the winter in six small-station groups [*Gunderson and Nelson*, 1965*b*]. A new inventory, called the opinion survey, was constructed from

TABLE 5. Numbers of Significant Predictors by Predictor Source, Occupational Subgroup, and Criterion

Subgroup and Criterion	Personal History and Hobbies	Opinion Survey Items	Firo-B Items	Friend Description Items	Psychiatric Ratings	Row Total
Seabees						
Emotion	8	6	1	...	...	15
Task	...	6	...	2	4	12
Social	2	11	1	1	...	15
Leadership	8	12	...	3	10	33
Technical personnel						
Emotion	4	3	3	3	...	13
Task	4	...	2	2	3	11
Social	4	4	2	3	1	14
Leadership	2	...	3	...	3	8
Scientists						
Emotion	7	2	11	4	3	27
Task	10	2	10	2	...	24
Social	9	4	14	6	2	35
Leadership	6	4	8	2	1	21

items included in the 10 earlier scales. This inventory was administered to new groups at three antarctic stations twice during the winter. The results corroborated earlier findings that scales reflecting motivation and group effectiveness tended to show consistent change in the direction of deterioration.

Intercorrelations among the 10 item clusters suggested the presence of a few major factors. As was previously mentioned, factor analyses of the earlier attitude study and group behavior description inventories were undertaken to determine the underlying factor structure of this attitude domain. Responses from two administrations of each inventory in each of two expeditions were factor-analyzed, this process producing eight separate analyses and replicating results from each inventory over samples and situations [*Shears and Gunderson,* 1966].

In all administrations of the attitude study inventory a large factor appeared that reflected motivation for and satisfaction with the antarctic assignment. Additional significant factors, such as feelings of job importance, emerged in one or another of the four administrations, but none were present consistently. In both first administrations of the group behavior description inventory, two similar major factors appeared, one encompassing the group compatibility and teamwork concept and the other encompassing group achievement and efficiency. At the second administration (end of winter) in both years, however, these two factors merged into one very large factor, two or three small additional factors absorbing the remaining content. Further analyses were required to establish the stability of the group compatibility and group accomplishment factors.

The new opinion survey inventory was administered to three stations on two occasions during the winter, and responses were factor-analyzed for each occasion. The first and largest factor in both administrations was that pertaining to group compatibility, teamwork, and pride in the group. This factor is essentially the same as the major factor previously obtained and leaves little doubt that liking for one's associates, cooperation, and esprit de corps are very important aspects of antarctic station effectiveness as perceived by the participants. The internal consistency reliability of this scale, based on item intercorrelations, was estimated to be about 0.88.

The second factor to emerge in both analyses was concerned with motivation for and personal satisfaction with the antarctic assignment. This factor also had been anticipated from results of earlier studies. The internal consistency reliability of this scale, based on item intercorrelations, was estimated to be about 0.83.

The third factor common to the two analyses was concerned predominantly with perceptions of the achievement of station members as a group. This factor was labeled group accomplishment. The internal consistency reliability of this scale was estimated to be about 0.75.

The simplicity and economy of the factor structures attained and the consistency from one occasion to another were encouraging and appeared to reflect the appropriateness of the questionnaire content for antarctic participants. The three principal factors identified, group compatibility, job morale, and group accomplishment, manifestly represent important aspects of individual and group life and are highly relevant to the goals of the antarctic program. Additional scales, labeled job importance and egalitarian atmosphere, were retained in the test battery because they seem to represent meaningful and distinct psychological concepts.

In earlier studies, by using the 10 item clusters, attitude changes were studied in nine subgroups from three antarctic expeditions. Of the attitude scales the group compatibility and group accomplishment scales most frequently showed significant changes from early winter or midwinter to the end of winter. Six of the nine subgroups showed significant changes in mean scores on these two scales. These results suggest that there was general agreement among station members at most stations that group harmony and efficiency declined during the winter months. Changes on the job morale and job importance scales were less marked and less consistent.

Measurement of attitude changes has been carried out during several recent expeditions. Results are considered in two parts: (1) changes in job satisfaction as measured by the job morale and job importance scales [*Ryman and Gunderson,* 1970] and (2) changes in group effectiveness as measured by the group compatibility and group accomplishment scales [*Doll and Gunderson,* 1971*a*].

Job satisfaction measures were administered during routine psychiatric screening before deployment and again on two occasions during the antarctic winter. The contents of the screening and the antarctic job attitude scales are shown in Table 6. Most of the items in the corresponding screening and antarctic measures are virtually identical in content except for the necessary modifications due to the test situations; the remaining items correlate very highly with the scale as a whole or have high loadings on their respective factors. The screening version of the first scale was labeled motivation, but the antarctic version was labeled job morale to be consistent with earlier usage. Both the screening version and the antarctic version of the second scale were labeled job importance.

The following relationships were examined in the study and are summarized here: (1) the relationships

TABLE 6. Composition of Screening and Antarctic Scales

Screening	Antarctic
Job Motivation/Morale	
1. Being part of an antarctic expedition will be the highlight of my career.	1. Being part of an antarctic expedition will be the highlight of my career.
2. I would like to stay in the Antarctic longer than what is now planned.	2. I wish that I could stay in the Antarctic longer than what is now planned.
3. Most of the people who go to the Antarctic will probably wish that they had stayed in the United States.	3. I often wish that I had never come to the Antarctic.
4. A large proportion of the people that I know would like to go to the Antarctic.	4. I would like to go on another antarctic expedition in the future.
5. I like the idea of waiting several months before we go to the Antarctic.	5. I am happier with my job here than I was in my last assignment.
	6. Time passes too slowly in the Antarctic.
Job Importance	
1. The knowledge and experience that I gain from an Antarctic expedition will help me advance in my career.	1. The knowledge and experience that I gain from this expedition will help me advance in my career.
2. In the Antarctic I will contribute more for the United States than I would at most other assignments.	2. In the Antarctic I am contributing more for the United States than I would be at most other assignments.
3. I think that I can contribute more on this expedition than most people can.	3. I feel that I am contributing as much on this expedition as others are.
4. I believe that I can make or help others make valuable scientific findings in the Antarctic.	4. The success or failure of this station depends as much on my job as it does on any other.
5. I would advise a friend who was considering joining a polar expedition to apply for the type of job that I will have in the Antarctic.	5. My job here is important enough to justify my spending all this time in the Antarctic.
	6. My present duties employ my abilities in the best possible way.

between screening measures of job attitude and later antarctic measures, (2) the test-retest stability of job morale and job importance measures during the winter, (3) the relationships of occupational subgroup and station size to job morale and job importance scores, and (4) the magnitude of changes in job morale and job importance during the winter months. Analyses were conducted separately for the three major occupational subgroups: Navy construction personnel, Navy technical-administrative personnel, and civilian scientists and technicians.

Motivation (screening) scores were substantially correlated with job morale scores in both early winter and late winter for the scientist subgroup ($p < 0.001$). Motivation was significantly correlated ($p < 0.05$) with job morale in early winter but not in late winter for the Navy technical-administrative subgroup. The same scales were not significantly correlated at either time period for the Navy construction subgroup.

Correlations between the screening version and the antarctic version for the job importance scale generally were lower than those for motivation and job morale. The screening measure correlated significantly with the antarctic measure for job importance in late winter but not in early winter for the scientist subgroup ($p < 0.05$) and in early winter but not in late winter for the Navy technical-administrative subgroup.

Test-retest correlations over the winter months were very high for the civilian subgroup on both scales (0.79 for job morale and 0.72 for job importance). Test-retest coefficients for the technical-administrative subgroup were 0.58 and 0.65, respectively, and, for the construction subgroup, 0.51 and 0.53, respectively. Thus the degree of stability in job attitudes during 7 months of isolation was substantial for all three occupational subgroups; however, the test-retest stability was much higher for the civilian subgroup than for the construction subgroup.

Job morale and job importance scores for Navy and civilian subgroups in early and late winter are shown in Table 7. Navy personnel as a whole had lower job morale scores than civilian personnel did in both early winter and late winter ($p < 0.001$). The difference was more pronounced at the end of winter. Job importance scores did not differ significantly between Navy personnel and civilian personnel in either time period.

When Navy personnel at large stations and those at small stations were compared, the difference was significant ($p < 0.05$) only for job importance in early winter; large-station personnel perceived their jobs as being more important than small-station personnel.

For the comparisons of large-station and small-station civilian personnel, only the difference on job morale in early winter was significant ($p < 0.01$); small-station civilians had higher satisfaction at that time than large-

TABLE 7. Job Attitude Mean Scores by Occupational Subgroup, Station Size, and Time Period

Station	Job Morale		Job Importance	
	Early Winter	Late Winter	Early Winter	Late Winter
Navy				
Large				
Mean	15.32	12.87	23.44	20.97
Standard deviation	5.99	6.13	5.13	5.79
Number	133	127	134	127
Small				
Mean	15.00	10.00	20.73	20.29
Standard deviation	7.54	6.55	5.17	5.39
Number	22	24	22	24
Civilian				
Large				
Mean	18.32	18.51	21.94	21.24
Standard deviation	5.63	6.65	4.81	4.73
Number	88	71	86	72
Small				
Mean	21.18	19.08	23.94	22.08
Standard deviation	4.33	6.99	3.49	6.95
Number	17	13	17	13

The attitude scales are scored so that high scores indicate high job morale or high job importance.

station civilians. Job morale scores were significantly higher for the civilian subgroup than for the Navy subgroup at both large stations and small stations and at both time periods.

At large stations the Navy subgroups scored higher than civilians on job importance in early winter ($p < 0.05$), but the difference between the two subgroups was negligible at the beginning of winter in that civilians scored higher on job importance than Navy personnel ($p < 0.05$). The difference at the end of winter was not significant.

The significance of changes in job morale and job importance during the winter was tested for four subgroups of personnel: large-station Navy, large-station civilian, small-station Navy, and small-station civilian. The difference for the Navy large-station subgroup on job morale was highly significant ($p < 0.001$), this significance indicating some deterioration in personal satisfaction for these Navy men during the winter. There was no change in job morale for civilians at large stations for early to late winter. In similar fashion, feelings of job importance declined significantly during the winter for Navy personnel at large stations ($p < 0.001$), but no change was recorded for civilian personnel. At small stations, Navy personnel again showed a significant decline in job morale ($p < 0.001$), but no similar change in job importance was observed. Relatively few civilians at small stations were included in the analysis, and the slight changes recorded in job morale and job importance did not approach statistical significance.

In summary the results were clear on a number of points. There was relatively high consistency in job satisfaction measures throughout the 18 months of involvement for the civilian subgroup, but consistency was much less for the Navy personnel; in fact, for the Navy subgroup there was no relationship between measures at screening and measures at the end of the antarctic period. Job satisfaction therefore was quite predictable from predeployment attitudes of the civilians but only slightly, if at all, predictable from those of the Navy subgroup.

Overall, the most positive attitudes were expressed by civilian personnel at small stations, and the most negative attitudes were expressed by Navy men at small stations. Thus there appeared to be some interaction between occupational subgroup and station size in the determination of attitudes, although this effect was not pronounced.

Navy men showed significant deterioration in morale or satisfaction through the winter regardless of station size, whereas civilians reported little or no change. It was concluded that occupational role was an important determinant of job satisfaction under conditions of long-term isolation and confinement.

Similar studies were conducted of station members' perceptions of group effectiveness as measured by the group compatibility and the group accomplishment scales. The data were from 19 subgroups during 1964–1968. Station groups were divided into Navy components and civilian components and into large size (20–30 members) and small size (8–11 members). Questionnaires containing the group compatibility and group accomplishment scales were administered in early winter (March) and late winter (September).

The compositions of the two scales were as follows.

Group compatibility. The men at this station work well together as a team. Group spirit at this station is high. Members of this station disagree a lot with one another. The men at this station are the kind of men that I like to spend a lot of time with. Everybody pulls together to get things done around here. There is a pretty good feeling among the men at this station.

Group accomplishment. The group does not seem to accomplish much. We usually have a good idea of what everyone else is doing. Everyone here would feel bad if the group did not accomplish its mission. When the going gets rough, this group is at its best. Everyone takes a lot of pride in what this group accomplishes. This group could achieve almost anything it sets out to do.

Group compatibility and group accomplishment scores were significantly lower at small stations than at large stations during both early winter and late winter. Examination of mean scores for the Navy subgroup and the civilian subgroup indicated clearly that these differences were largely accounted for by the Navy subgroup. For both early winter and late winter administrations the differences between large-station responses and small-station responses were significant on the two scales for Navy personnel but not for civilian personnel. The differences were most pronounced in late winter. The change in group compatibility scores from early to late winter was especially marked for the Navy personnel. In early winter, civilian personnel had somewhat less favorable perceptions of group compatibility and group accomplishment than the Navy personnel at both large stations and small stations. At the end of winter, however, small-station Navy personnel tended to have more unfavorable attitudes on both scales than any other subgroup. The size of station thus was directly related to perceptions of compatibility and achievement among military personnel. Although similar trends were apparent for civilian members, in no instances did these differences approach statistical significance.

EMOTIONAL SYMPTOMS

The types and intensities of emotional distress experienced in the antarctic environment are clearly pertinent to the assessment of individual adaptation. Studies concerned with the measurement of emotional symptoms have been conducted since the IGY and will be reviewed in this section.

It has long been known that reductions in the intensity or variety of external stimulation from normal levels may have profound effects on thought and behavior. *Hebb* [1949] in his experimental work on what he called 'perceptual isolation' suggested that monotonous sensory stimulation produces a disruption of the capacity to learn or even to think. *Kubzansky and Leiderman* [1961] in a critique of experimental work on sensory deprivation stated that studies thus far indicate that the absence of stimulation makes the individual less efficient and induces strong affective states associated with marked changes in motivation. Varied environmental stimulation, then, seems to be vital in maintaining the efficiency and stability of behavior.

Questionnaires indicating the presence and intensity of a variety of common somatic and emotional symptoms were administered to a number of antarctic groups during both years of the IGY. Tests were administered at the beginning of winter, at midwinter, and at the end of winter. The most prevalent symptoms in midwinter in both expeditions were sleep disturbances (difficulty in falling asleep or staying asleep, waking up at night, and feeling tired during the day) and depression (feeling blue and feeling lonely). Headaches, easy annoyance or irritation, and soreness of muscles also were reported frequently both years.

In the first expedition, significant increases in severity from prewinter to midwinter were noted for the following symptoms: having difficulty in falling asleep or staying asleep, waking up at night, having bad dreams, feeling blue, feeling lonely, feeling annoyed or irritated, feeling critical of others, having headaches, feeling tired during the day, and having pains in the lower back. In the second expedition the most marked changes occurred in the same items that differed significantly from prewinter to midwinter in the first expedition, that is, items pertaining to sleep disturbances, depression, and irritability. It seems clear that emotional disturbances and somatic complaints tended to increase during the winter months. These symptoms appeared psychologically unpleasant and presumably would have a deleterious effect on work effectiveness and social adjustment. This assumption was supported by significant correlations between symptom scales (clusters of intercorrelated items labeled anxiety, depression, aggression, and insomnia) and attitude scales (job morale, job importance, and group compatibility) administered near the end of winter during one IGY expedition. Increases in anxiety, depression, or irritability among members of small isolated groups would seem to present a definite threat to group harmony and effectiveness.

A more recent study covering 1964–1968 investigated the relationship of symptom scores to occupational role

(Navy versus civilian) and station size [*Doll and Gunderson,* 1971*b*]. Symptom questionnaires were administered twice during the year, in March and September, and scores on four scales were derived at each administration. The scales and their items were as follows: depression, feeling blue and feeling lonely; insomnia, having difficulty in falling asleep or staying asleep, waking up at night, and feeling tired during the day; hostility, feeling easily annoyed or irritated and feeling critical of others; and anxiety, feeling nervous and tense, being unable to concentrate, and feeling uneasy or worried.

Depression scores were higher for the Navy subgroup than for the civilians in both early winter and late winter ($p < 0.001$). The difference in insomnia scores between the two subgroups was negligible in early winter but highly significant at the end of winter ($p < 0.001$), Navy men reporting more sleep disturbances. No differences were present on the anxiety scale for either administration. The civilian subgroup scored higher than the Navy subgroup on the hostility scale in early winter, but no difference was evident at the end of winter.

HOBBY AND LEISURE INTERESTS

Hobbies and leisure activities of antarctic participants have been of increasing interest as an area of psychological research. During the winter months, when work activities are reduced for most station members, free time is available for avocational and recreational activities. The use made of this time and the satisfaction or boredom experienced by the individual may significantly affect his psychological adjustment. An investigation was made of hobby interest likes and dislikes and leisure activities preferred by wintering-over personnel at small stations during 1964–1968 [*Doll and Gunderson,* 1969]. Analyses were conducted separately for Navy construction, Navy technical-administrative, and civilian personnel. Included in the psychiatric screening program was an inventory designed to elicit attitudes toward the following hobbies: building models, ham radio, classical music, motorcycles, plays and dramas, popular music, painting and drawing, individual sports, reading magazines, country-western music, collecting stamps, hunting and fishing, team sports, repairing things, reading books, hiking and camping, movies, hot rods, photography, and cards.

In addition to the hobby questionnaire given at screening, another questionnaire describing leisure activity behavior was administered twice during the winter in Antarctica. This questionnaire contained the following 20 items: movies, bull session (present job), bull session (past job), bull session (general), reading fiction, reading biography, reading religious literature, reading technical magazines, studying courses, ham radio, writing letters, physical exercise, painting and drawing, 'happy hour,' cards, chess or checkers, pool or billiards, classical music, popular music, and country-western music.

The activity questionnaire items were answered on a 5-point scale anchored at one end by 'not at all' and at the other end by 'every day,' and they referred to leisure activities engaged in during the previous 3 months.

The hobby items were answered on a 3-point scale, namely, 'like strongly,' 'neither like nor dislike,' and 'dislike strongly,' and scored 3, 2, and 1, respectively.

Pearson product-moment correlations between the hobby items and the activity items were computed separately for the three occupational subgroups. Correlations were computed also between the early administration and the late administration of the leisure activity questionnaire items.

The relative attractiveness of the various hobbies at the time of screening was compared for the three occupational subgroups. Only hiking and camping were considered by all three occupational subgroups as being highly desirable. Both the Seabees and the technical-administrative subgroup were attracted to reading magazines and to team sports. The Seabees and the civilians shared a high degree of interest in repairing things, whereas both the technical-administrative subgroup and the civilians were attracted to individual sports and reading books. Each of the occupational subgroups was highly interested in one hobby not shared by the other two groups: for the Seabee this was hunting and fishing, for the technical-administrative subgroup it was popular music, and for the civilians it was classical music.

There was agreement by all three subgroups that collecting stamps and painting and drawing were least desirable. In addition to these two hobbies, the Seabees and the technical-administrative subgroup agreed on the relative undesirability of working with radio equipment. The technical-administrative personnel and the civilians shared a dislike for hot rods and motorcycles. The Seabees were unique in having a strong distaste for plays and dramas, painting and drawing, and classical music. Civilians, on the other hand, were unique in their rather extreme dislike for country-western music.

There was considerable overlap among the three occupational subgroups in the leisure activities preferred or rejected during the winter months in Antarctica. All three subgroups spent much of their leisure time attending movies, discussing their present job, and listening to popular music. The Seabees and the technical-administrative subgroup shared an interest in country-western music and in discussions of their past jobs. The civilians, on the other hand, were distinct from the other two subgroups in that they spent a relatively high proportion of their leisure time listening to classical music and reading technical magazines.

There was even higher agreement among the three subgroups regarding the least preferred activities for the early winter administration. Reading biographies, playing chess or checkers, reading religious literature, and painting and drawing were avoided by all three subgroups. The technical-administrative subgroup and the civilians did not like to play cards. The Seabees were unique in their avoidance of physical exercise as a leisure activity.

There did not appear to be any striking changes in leisure activities from early to late winter. There was an increase in physical exercise and in fiction reading; at the same time there was a sharp decrease in the popularity of classical music with the Navy subgroup. In general, the most popular activities in early winter retained their popularity throughout the winter.

PSYCHOLOGICAL STUDIES AT AUSTRALIAN STATIONS

A series of four reports by *Owens* [1966, 1967*a*, *b*, 1968] describes the development of performance criteria and the relationships of personality and biographical variables to these criteria in wintering-over parties of the Australian National Antarctic Research Expedition. Four stations were involved in the studies, and the station groups ranged in size from 7 to 32.

The first report describes the construction of scales on which station leaders rated their men and a factor analysis of the ratings for 135 men in eight wintering parties. Principal axis factor analysis and a varimax rotation yielded four distinct and interpretable factors closely resembling the results obtained independently by *Gunderson and Nelson* [1965*a*] at U.S. stations. A comparison of the similar sets of behavior factors is shown in Table 8. The most salient rating items for defining each factor are indicated.

At the Australian stations, two distinct environmental situations were differentiated: (1) living at a permanent base (relative safety and comfort with low stimulus variation) and (2) participating in field parties (high hazard and high stimulation). The field trip factor obtained in the factor analysis was interpreted as a motivational variable related to the specific situation at Australian stations. It was suggested that selection for overall suitability would require consideration of expected participation in both situations.

TABLE 8. Comparison of Supervisor Behavior Factors from U.S. and Australian Stations

Behavior Factor	Salient Rating Items
Australian Study	
Task factor	Attention to duty, reliability, technical proficiency
Interaction factor	Social value, consideration for others, willingness to assist others
Intrapersonal factor (stress tolerance)	Behavior under stress, judgment
U.S. Studies	
Task motivation	Motivation, industriousness, proficiency
Social compatibility	Likability, consideration for others, cheerfulness
Emotional stability	Emotional control, acceptance of authority

The field task proficiency factor, which appeared in the Australian study only, is omitted.

The general criterion item, Would you have him again?, rated on a 5-point scale, weighed most highly on the interpersonal factor, next on the task factor, and lowest on the intrapersonal factor.

In the second and third reports in this series, relationships between the performance criteria described above and personality attributes as measured by the T. Leary interpersonal checklist (ICL) were analyzed. In part 2, ICL scores reflecting 10 personality dimensions originally suggested by the test authors were correlated with the 12 performance scales. All ratings were made by a single leader at each station. The love score on the ICL was related to the intrapersonal performance factor, whereas the dominance score was correlated with both the task dimension and the interpersonal dimension. Other personality variables also correlated with the performance criteria, these correlations providing some indication of valued behavior traits in the antarctic environment.

In the third part of the study the items of the ICL were factor-analyzed, and the first 10 factors were subjected to rotation. Four major and five minor factors emerged; the major factors were identified as agreeableness, leadership, rigid defensiveness, and group cooperation.

On the basis of their relationships with the task, interpersonal, and intrapersonal criteria, the following desirable personality attributes (ICL dimensions) for antarctic service were indicated: agreeableness, leadership, self-reliance, group cooperation, and openness. The undesirable qualities were rigid defensiveness and dependence on others for task direction.

Using the overall performance rating 'want him again'

and the 'value on a field trip' item as criteria, a number of biographical variables were studied as possible indicators of antarctic adjustment.

Age was not related to the overall criterion. Age was negatively correlated with the field trip criterion; as was indicated previously, the field trip situation was a much more physically demanding one than the living on base situation.

Marital status was found to be related to the overall criterion; married men received poorer performance ratings than single men. Because age did not account for this difference, it appeared that marriage might have affected the individual's emotional adjustment during prolonged separation from home. Performance ratings of married men were even less favorable on the field trip criterion than on the overall criterion.

Occupational specialty, education, family stability, and birth order all were unrelated to overall performance ratings. Previous antarctic experience was not related to overall performance but was positively related to the field trip criterion. When the relationship between previous experience and overall performance was examined separately for each occupational specialty, it was found that experience was positively related to performance for radio operators and meteorological technicians but not for scientists and maintenance personnel.

Owens concluded that the relationships between biographical characteristics and performance criteria were moderated by task or occupation. Similar findings have been reported in U.S. studies.

PSYCHOLOGICAL STUDIES AT NEW ZEALAND STATIONS

Scott base was established by New Zealand in 1956 to serve as the terminal for the Commonwealth Trans-Antarctic Expedition and as a scientific laboratory during the IGY. This station is a short distance from the large U.S. base at McMurdo Sound. Psychological data were first collected at Scott base by A. J. W. Taylor in 1967. Members of two successive expeditions were interviewed at the end of winter isolation. *Taylor* [1969] reported the typical responses of station members to a standardized set of questions concerned with demographic and employment backgrounds, reasons for volunteering, goals and achievements during the year, perceptions of time, adjustment problems including sleep disturbances, closest associates, group cohesiveness, and improvement of selection procedures.

The most frequent motives given for volunteering were: saving money, accepting a difficult challenge, and interest in or curiosity about the Antarctic. Wives and children or female companionship were missed most, along with the variety and novelty encountered in the normal home environment. Open expression of aggression tended to be avoided, but aggression was expressed indirectly through arguments about politics and similar issues. Most of the members of both parties experienced sleep disturbances, in some cases over periods of several weeks.

A second phase of the study involved a comparison of the personality attributes of three wintering parties as measured by the R. B. Cattell 16 personality factor test. The first group was tested at the end of the year, the second group was tested at both the beginning of the year and the end, and the third group was tested at the beginning of the year. Significant variability among the three groups was present only for the practical (conventional)–imaginative scale. This single significant difference could well have occurred by chance. In comparison with U.S. norms for these scales, all three parties appeared to be more reserved, intelligent, shy, trusting, self-sufficient, and controlled than the population at large. Extroversion was not characteristic of these antarctic groups.

In a subsequent paper, *Taylor and Shurley* [1971] compared 16 personality factor test scores for the Scott base group tested before and after winter isolation with a South Pole station group. Before winter the South Pole group described themselves as being more sober and emotionally stable but less conservative than the Scott group. The South Pole group was more controlled and composed than the Scott group at the end of winter. The South Pole personnel were more self-sufficient both before winter and after winter than the Scott group. During the year the Scott base personnel indicated that they became more sober (less happy-go-lucky), whereas the South Pole men indicated that they became more self-sufficient and controlled. The author speculated that the extreme degree of isolation and confinement experienced at South Pole might explain the trend toward greater self-sufficiency observed in that setting.

Acknowledgment. Report 71-14, supported by the U.S. Bureau of Medicine and Surgery, Department of the Navy, under research work unit MF12.524.001-9003D. Opinions expressed are those of the author and are not to be construed as necessarily reflecting the official view or endorsement of the Department of the Navy.

REFERENCES

Allport, G. W., P. E. Vernon, and G. Lindzey, *A Study of Values,* revised ed., Houghton Mifflin, Cambridge, Mass., 1951.

Borgatta, E. F., The structure of personality characteristics, *Behav. Sci., 9,* 8–17, 1964.

Campbell, D. T., and D. W. Fiske, Convergent and discriminant validation by the multitrait-multimethod matrix, *Psychol. Bull., 56,* 81–105, 1959.

Doll, R. E., and E. K. E. Gunderson, Hobby interest and leisure activity behavior among station members in Antarctica, *Unit Rep. 69-31,* U.S. Navy Med. Neuropsychiat. Res. Unit, San Diego, Calif., 1969.

Doll, R. E., and E. K. E. Gunderson, The influence of group size on perceived compatibility and achievement in an extreme environment, *Personnel Psychol., 24,* 305–310, 1971*a*.

Doll, R. E., and E. K. E. Gunderson, Group size, occupational status, and psychological symptomatology in an extreme environment, *J. Clin. Psychol., 27,* 196–198, 1971*b*.

Doll, R. E., E. K. E. Gunderson, and D. H. Ryman, Relative predictability of occupational groups and performance criteria in an extreme environment, *J. Clin. Psychol., 25,* 399–402, 1969.

Edwards, A. L., *Manual for the Edwards Personal Preference Schedule,* revised ed., Psychological Corporation, New York, 1959.

Gordon, L. V., *Manual for Survey of Interpersonal Values,* Science Research Association, Chicago, Ill., 1960.

Gunderson, E. K. E., Emotional symptoms in extremely isolated groups, *Arch. Gen. Psychiat., 9,* 362–368, 1963.

Gunderson, E. K. E., The reliability of personality ratings under varied assessment conditions, *J. Clin. Psychol., 21,* 161–164, 1965*a*.

Gunderson, E. K. E., Determinants of reliability in personality ratings, *J. Clin. Psychol., 21,* 164–169, 1965*b*.

Gunderson, E. K. E., and E. L. Kapfer, Variability in factor structures of clinicians' personality ratings, *J. Couns. Psychol., 30,* 366, 1966*a*.

Gunderson, E. K. E., and E. L. Kapfer, The predictive validity of clinical ratings for an extreme environment, *Brit. J. Psychiat., 112,* 405–412, 1966*b*.

Gunderson, E. K. E., and J. L. Mahan, Cultural and psychological differences among occupational groups, *J. Psychol., 62,* 287–304, 1966.

Gunderson, E. K. E., and P. D. Nelson, Clinical agreement in assessing for an unknown environment, *J. Clin. Psychol., 20,* 290–295, 1964.

Gunderson, E. K. E., and P. D. Nelson, Biographical predictors of performance in an extreme environment, *J. Psychol., 61,* 59–67, 1965*a*.

Gunderson, E. K. E., and P. D. Nelson, Measurement of group effectiveness in natural isolated groups, *J. Soc. Psychol., 66,* 241–249, 1965*b*.

Gunderson, E. K. E., and P. D. Nelson, Criterion measures for extremely isolated groups, *Personnel Psychol., 19,* 67–80, 1966.

Gunderson, E. K. E., and D. H. Ryman, Convergent and discriminant validities of performance evaluations in extremely isolated groups, *Personnel Psychol., 24,* 715–724, 1971.

Hebb, D. O., *The Organization of Behavior,* John Wiley, New York, 1949.

Kubzansky, P. D., and P. H. Leiderman, Sensory deprivation: An overview, in *Sensory Deprivation,* edited by P. Soloman et al., pp. 221–238, Harvard University Press, Cambridge, Mass., 1961.

McGuire, F., and S. Tolchin, Group adjustment at the south pole, *J. Ment. Sci., 107,* 954–960, 1961.

Mullin, C. S., Some psychological aspects of isolated antarctic living, *Amer. J. Psychiat., 117,* 323–325, 1960.

Mullin, C. S., and H. J. Connery, Psychological studies at an antarctic IGY station, *U.S. Armed Forces Med. J., 10,* 290–292, 1959.

Nardini, J. E., R. S. Herrmann, and J. E. Rasmussen, Navy psychiatric assessment in the Antarctic, *Amer. J. Psychiat., 119,* 97–105, 1962.

Owens, A. G., The assessment of individual performance in small antarctic groups, 1, OIC's rating scales, research report, Psychol. Res. Unit, Aust. Mil. Forces, Melbourne, Australia, Aug. 1966.

Owens, A. G., The assessment of individual performance in small antarctic groups, 2, Ratings on the Leary interpersonal checklist, research report, Psychol. Res. Unit, Aust. Mil. Forces, Melbourne, Australia, Aug. 1967*a*.

Owens, A. G., The assessment of performance in small antarctic groups, 3, Factor analysis of Leary interpersonal checklist items, research report, Psychol. Res. Unit, Aust. Mil. Forces, Melbourne, Australia, Oct. 1967*b*.

Owens, A. G., Some biographical correlates of assessed performance in small antarctic groups, research report, Psychol. Res. Unit, Aust. Mil. Forces, Melbourne, Australia, Oct. 1968.

Palmai, G., Psychological aspects of transient populations in Antarctica, Medicine and Public Health in the Arctic and Antarctic, *Public Health Pap. 18,* pp. 146–158, World Health Organ., Geneva, Switzerland, 1963*a*.

Palmai, G., Psychological observations on an isolated group in Antarctica, *Brit. J. Psychiat., 109,* 364–370, 1963*b*.

Radloff, R., and R. Helmreich, *Groups under Stress: Psychological Research in SEALAB 2,* Appleton-Century-Crofts, New York, 1968.

Rohrer, J. H., Interpersonal relationships in isolated small groups, in *Psychophysiological Aspects of Space Flight,* edited by B. E. Flaherty, pp. 263–271, Columbia University Press, New York, 1961.

Ryman, D. H., and E. K. E. Gunderson, Factors affecting the stability of job attitudes in long-term isolated groups, *Unit Rep. 70-13,* U.S. Navy Med. Neuropsychiat. Res. Unit, San Diego, Calif., 1970.

Schutz, W. C., *Firo: A Three-Dimensional Theory of Interpersonal Behavior,* Holt, Rinehart, and Winston, New York, 1958.

Shears, L. M., and E. K. E. Gunderson, Stable attitude factors in natural isolated groups, *J. Soc. Psychol., 70,* 199–204, 1966.

Smith, W. M., Scientific personnel in Antarctica: Their recruitment, selection, and performance, *Psychol. Rep., 9,* 163–182, 1961.

Smith, W. M., Observations over the lifetime of a small isolated group: Structure, danger, boredom, and vision, *Psychol. Rep., 19,* 475–514, 1966.

Smith, W. M., and M. B. Jones, Astronauts, antarctic scientists and personal autonomy, *J. Aerosp. Med., 33,* 162–166, 1962.

Taylor, A. J. W., Ability, stability, and social adjustment among Scott base personnel, Antarctica, *Occup. Psychol., 43,* 81–93, 1969.

Taylor, A. J. W., and J. T. Shurley, Some antarctic troglodytes, *Int. Rev. Appl. Psychol., 20,* 143–148, 1971.